THE PICKERING MASTERS

THE WORKS OF
CHARLES DARWIN

Volume 26. *The different forms of flowers on plants of
the same species*

THE WORKS OF
CHARLES DARWIN

EDITED BY

PAUL H. BARRETT & R. B. FREEMAN

ADVISOR: PETER GAUTREY

VOLUME
26

THE DIFFERENT FORMS
OF FLOWERS ON
PLANTS OF THE SAME SPECIES

PREFACE BY FRANCIS DARWIN

Routledge
Taylor & Francis Group

LONDON

First published 1992 by Pickering & Chatto (Publishers) Limited

Published 2016 by Routledge
2 Park Square, Milton Park, Abingdon, Oxon OX14 4RN

Routledge is an imprint of the Taylor & Francis Group, an informa business

British Library Cataloguing in Publication Data
Darwin, Charles, *1809–1882*
  The works of Charles Darwin.
  Vols. 21–29
  I. Title II. Barrett, Paul H. (Paul Howard), *1917–1987*
  III. Freeman, R. B. (Richard Brooke), *1915–1986*
  IIII. Gautrey, Peter, *1925–*
  575
  ISBN 13: 978-1-851-96406-2 (hbk)

INTRODUCTION TO VOLUME TWENTY-SIX

The Different Forms of Flowers on plants of the Same Species. [Second edition] Preface by Francis Darwin, 1884. Freeman 1281.

The first postulate of the theory of Natural Selection is, 'There is variation among the individuals of a species'. This book is full of facts and figures supporting this postulate. Darwin gathered data on 144 plant species and another 40 plant genera. Much of his data were collected by conducting innumerable breeding experiments. His goals were to discover reproductive advantage of various flower structures, e.g., varying lengths of pistils and stamens in members of the same species. His measurements of survival advantage were numbers of seeds produced, numbers of fertile seeds, numbers of pollen grains, and 'prepotency' of grains.

His studies demonstrate the intergradation of flower forms from hermaphroditic through varying degrees of polymorphism to monoecious and to dioecious types. Thus in some instances both male and female parts are present in the same flower, and in others, male and female parts are in separate flowers, with varying degrees of mixtures of each in certain forms.

The contents of this book had previously been published in five articles in *Journal of Proceedings of the Linnean Society*, 1862 to 1868.

The second edition published in 1884 has been chosen for this collection because Francis Darwin added a preface summarizing new contributions to the literature.

*The different forms
of flowers on plants
of the same species*

THE

DIFFERENT FORMS OF FLOWERS

ON

PLANTS OF THE SAME SPECIES.

By CHARLES DARWIN, M.A., F.R.S.

THIRD THOUSAND.

WITH ILLUSTRATIONS.

LONDON:
JOHN MURRAY, ALBEMARLE STREET.
1884.

The right of Translation is reserved.

TO

PROFESSOR ASA GRAY

THIS VOLUME IS DEDICATED

BY THE AUTHOR

AS A SMALL TRIBUTE OF RESPECT AND

AFFECTION

PREFACE
TO THE REPRINT OF 1884

The text of the second edition has been left untouched, and I have merely given an account (which does not pretend to be complete) of the progress of the subject since 1880.

HETEROSTYLED PLANTS

C. E. Bessey (*American Naturalist*, June, 1880, p. 417) made careful measurements of the corolla, stamens, and style in a number of flowers of *Lithospermum longiflorum*. He shows that the length of corolla, and especially the length of the style, is very variable. An appearance of dimorphism is thus produced; but measurements of the pollen show that there is no real heterostylism.

C. B. Clarke (*Journ. Linn. Soc.*, xvii, p. 159) has made the curious observation that in *Adenosacme longifolia*, the difference between the long- and short-styled forms amounts to what would usually be called a character of first-rate systematic importance. In the short-styled flowers, the stamens are on the corolla; in the long-styled, they are at its very base, almost free from it. In this form the corolla separates and leaves the stamens standing on the ovary. /

He also describes two forms of *Randia uliginosa*, (i) having large sessile flowers with separate stigmas and producing a large fruit; (ii) small pedunculated flowers with clavate stigmas, and producing smaller fruit.

C. B. Clarke (*Journ. Linn. Soc.*, xviii, p. 524) shows that Macrotomia is dimorphic like Arnebia. Mr Clarke mentions as one of the earliest good notices of heterostylism that Fischer and Meyer (*Enum. Pl. Schrenk.*, p. 34, published in 1841) speak of Macrotomia as having *specimina longistyla* and *brevistyla*.

Breitenbach (*Botanische Zeitung*, 1880, p. 577) believes that the ancestor of the heterostyled Primulae was homostyled. He grounds his belief on the examination of a large number of plants of *P. elatior*,

Jacq., and on some facts connected with the ontogenesis of the flowers. This opinion has been adversely criticized by W. Behrens (*Botanisches Centralblatt*, 1880, p. 1082) and by Hermann Müller (*Bot. Zeitung*, 1880, p. 733).

A. Ernst (Carácas) (*Nature*, xxi, 1880, p. 217) shows by measurement and experiment that *Melochia parviflora* is heterostyled (dimorphic).

According to J. Todd (*American Naturalist*, xv, 1881, p. 997), Black mustard (*B. nigra*) has two forms of flower, differing in the length of the pistil; the stamens being of approximately the same length in the two forms.

Trelease (*American Naturalist*, xvi, 1882, p. 13) describes two forms of *Oxalis violacea*, which appear to be long- and short-styled forms of a trimorphic species. No mid-styled flowers could be found, and Trelease is inclined to believe that the species is dimorphic.

Ig. Urban (*Sitz. Bot. Verein, Prov. Brandenburg,* / xxiv, 1882) states that the Turneraceae contain a large proportion of dimorphic plants. His monograph on this family I know only from the abstract in the *Botan. Centralblatt*, p. 207. He has made the following interesting observations: 'In the Turneraceae the dimorphic species tend to be perennials, with conspicuous flowers, while the monomorphic species have smaller flowers, and are chiefly annuals.' He states that a tendency to dimorphism in the monomorphic species expresses itself only in elongation of the style.

In the subjects kindred to those considered in chapter VII rather more work has been done.

F. Ludwig (*Zeitschrift f. d. gesam. Naturwiss.*, 1879, p. 44) describes three plant-forms in *Plantago lanceolata*.

1. Hermaphrodites with white anthers.

2. Semi-females, with small shrivelled yellow anthers, containing a small quantity of pollen, of which many grains are bad.

3. Purely female form.

Ludwig has noticed the tendency described by Delpino to entomophily in Plantago, the flowers being often fairly conspicuous, and are visited by insects.

Ludwig draws some interesting general conclusions on Gynodioecious plants.

1. They are all more or less dichogamic.

2. In the protandrous forms the females are more numerous at the beginning of the season. In the protogynous forms the reverse is the case.

3. Abortive anthers often degrade into perianth segments.

4. He confirms the received opinion that female flowers are smaller than hermaphrodites.

He discusses the origin of dioeciousness, assigning / the first rank in the chain of causes to dichogamy. Similar views are given in the present edition, p. 283, in connection with observations by Hildebrand.

In a subsequent paper (*Botan. Centralblatt*, 1880, iv, p. 829) he describes a similar gynodioecious condition in some Stellarias and Cerastiums. Here there are pure female, semi-female, and herma-phrodite plants, the flowers of the female form being smaller than those of the others. This distribution of sex he calls 'gynodimorphism', a condition which he describes (*Bot. Centralblatt*, 1880, p. 1021) as occurring in *Arenaria ciliata* and *Alsine verna*.

F. Ludwig (*Kosmos*,[1] 1880–1, viii, 357) describes two forms of *Erodium cicutarium*. The first, which is distinguished by having nectar-guides, is protandrous, and adapted for fertilization by insects. The second form is weakly protogynous and autogamic. This form has no nectar-guides, and the petals are usually shed during the day on which the flowers open. It resembles *E. moschatum*, which is homogamic (or weakly protogynous). The first form is more like *E. macrodenum*, which is markedly protandrous, and where autogamy is impossible.

Hermann Müller (*Nature*, xxiii, p. 337, 1881) has shown that *Syringa persica* is gynomonoecious, having on the same inflorescence a majority of hermaphrodite flowers of large size, and a minority of small-sized females.

Stellaria glauca and *Sherardia arvensis* are gynodioecious.

H. Müller has also written an important paper on *Centaurea jacea* (*Kosmos*, x and *Nature*, xxv) in which he publishes his change of opinion as to the origin of / gynodioeciousness. Three different forms occur, but on any given plant the flowers are of one kind. There is a normal hermaphrodite form, and two divergent forms which are practically male and female, and which are distinguished from the hermaphro-dite form by having conspicuous sexless ray-florets; of the two, the male flowers are the more conspicuous. The female florets have shrunken anthers devoid of pollen; the male florets have pistils which do not open, and are therefore functionless. Numerous gradational forms exist which render the whole case especially instructive, and it

---

[1] See also *Irmischia*, 1881, no. 1, and *Bot. Centralblatt*, xii, p. 83 and viii, p. 87.

was a study of these gradations which induced Müller to give up his theory of gynodioecious plants. Müller formerly explained the origin of gynodioeciousness by supposing that those flowers which are smaller and less conspicuous than the average tend to be visited last by insects, so that their pollen is useless. In Centaurea the reduction of anthers is found beginning in flower-heads which are *not* less conspicuous than the average. Müller therefore gives up his former theory and agrees with view proposed by my father.[a]

Potonié (*Sitz. d. Ges. naturforsch. Freunde zu Berlin*, 1880, p. 85, quoted in the *Bot. Zeitung*, 1880, p. 749) believes that in the gynodioecious *Salvia pratensis* the existence of a female form serves to ensure fertilization by a distinct plant.

But H. Müller (*Bot. Zeitung*, 1880, p. 749) shows that in the hermaphrodites, bees commonly visit the lower and temporarily female flowers before passing on to the upper male flowers, and that this ensures cross-fertilization between different plants.

Solms-Laubach (*Abhand. K. Gesell. Wiss. Göttingen*, xxviii, and *Kosmos*, 1881) has given in his / valuable work on caprification an account of the relation of the sexes in the cultivated fig and the caprificus.

HETERANTHY

The existence of different kinds of anthers in homostyled flowers is of interest as bearing on heterostylism.

F. Ludwig (*Bot. Centralblatt*, 1880, pp. 246 and 1210) gives an account of the heteranthy of *Plantago major*, of which two forms exist, one with brown, the other with yellow anthers; the latter plants are much rarer than the brown-anthered form. In another communication to the same journal (1880, p. 861), he describes the heteranthy of *Poterium sanguisorba*, and of a number of grasses, e.g. *Lolium dactylis*, *Festuca*, *Aira*.

F. Müller (*Nature*, xxiv, 1881, p. 307), has made the curious observation that in the Melastomaceous *Heeria*, *sp.*, there are two sets of anthers: (1) yellow ones serving as plunder to bees; (2) red ones so placed as to subserve cross-fertilization.

H. Müller (*Nature*, 1882, p. 30) showed that in *Tinnantia undata* (Commelynaceae), as in Heeria, two sets of anthers exist; one set which attract pollen-seeking insects, the other which cover the insect with

---

[a] A short paper by H. Müller on gynodioeciousness in the genus Dianthus, appeared in *Nature*, 1881, xxiv.

pollen. The upper stamens have yellow tufts of hair, which (as in Tradescantia) serve as supports for visiting insects. The pollen grains are smaller in the upper stamens. In *Commelyna coelestis* and *communis*, there is somewhat similar arrangement.

In a species of Melastoma, which has also two sets of stamens, H. O. Forbes (*Nature*, 1882, p. 386), saw bees going straight to the yellow stamens, i.e. to those which serve as an attraction. The yellow anthers have the smaller pollen grains, but those from the other set / of anthers were the only ones seen to exsert tubes in the stigmas.

J. E. Todd (*American Naturalist*, xvi, 1882, p. 281) gives a curious account of *Solanum rostratum*, in which the pollen for fertilization is the product of a single long-curved anther; while the four other anthers are small, and serve to supply pollen to the bees visiting the flower. The stigma is so placed that it receives pollen from the part of the bee dusted by the long anther.

CLEISTOGAMIC FLOWERS

According to P. Ascherson (*Bulletin Soc. Linn. de Paris*, 1880, p. 250),[3] *Helianthemum salicifolium* was shown by Linnaeus to produce ripe seed from closed flowers. Ascherson describes the cleistogamic flowers of *H. Kahiricum* and *H. Lippii γ. micranthum*, Boiss. Schweinfurth is given as authority for the existence of cleistogamic flowers in *Salvia lanigera*. The following species are said to be 'often cleistogamic': *Lamium amplexicaule, Juncus bufonius, Ajuga Iva, Campanula dimorphantha*.

In a second paper (*Sitz. d. Gesch. naturf. Freunde zu Berlin*, 1880, p. 97, quoted in the *Bot. Centralblatt*) Ascherson gives a further account of the cleistogamy of *Helianthemum Kahiricum*. The flowers are open in the early morning, so that cross-fertilization is possible; the petals fall off in the course of the day, and the sepals closely embrace the stamens and pistils, and thus convert the flower into a cleistogamic one.

Baron E. Eggers (*Bot. Centralblatt*, 1881, viii, p. 57) states that *Sinapis arvensis*, when grown in West Indies, produces cleistogamic flowers. /

The following Acanthaceae have cleistogamic flowers: *Stenandrium rupestre, Diclipetra assurgens, Stemonacanthus coccineus, Dianhera sessilis, Blechum Brownei*.

Among other families: *Erithalis fruticosa, (Rubiaceae), Polystachya luteola*, are also cleistogamic.

The curious flowers of *Pavonia hastata* are described by E. Heckel

[3] As abstracted in the *Bot. Centralblatt*.

(*Comptes rendus*, lxxxix, p. 609). This species has cleistogamic flowers, which chiefly differ in appearance from the perfect flowers, in having no nectar-guides; there are, as usual, no nectaries. The pollen is entomophilous in character, and it is said that the tubes are protruded while the pollen is in the anthers.

F. Ludwig (*Bot. Centralblatt*, 1880, p. 861) mentions *Plantago virginica* as producing under cultivation only cleistogamic flowers.

F. Müller (*Nature*, xix, 1879, p. 463) shows that the curious sub-merged Podostomaceae of Brazil produce flowers which are probably cleistogamic.

Solms Laubach (*Göttingen Nachrichten*, June, 1882) has written an interesting paper on Heteranthera, a plant belonging to the Ponteder-eaceae. He describes the cleistogamy of some species of the genus, and points out that the form and distribution of the cleistogamic flowers serves as a specific character, without which *H. callaefolia* could not be distinguished from *H. Kotschyana*.

January, 1884   FRANCIS DARWIN /

PREFACE
TO THE SECOND EDITION

Since the publication of the first edition of this book in 1877, several articles have appeared on the subjects therein discussed, and many letters have been received by me. I will here briefly state their nature, as an aid to any one who may afterwards pursue the same subjects. The text has been left as it originally appeared, excepting that a few errors have been corrected.

Dr A. Ernst has proved in the clearest manner (*Nature*, 1 January, 1880, p. 217), that *Melochia parvifolia*, which is a common plant near Caracas, is heterostyled. The pollen-grains differ in the usual manner in size in the two forms, as do the papillae on their stigmas. The illegitimate unions, especially when pollen from the same flower was employed, were much less fertile than the legitimate ones. A new family, the Byttneriaceae, is thus added to those including heterostyled plants.

Errara and Gevaert have published a paper on the heterostylism of *Primula elatior* in *Bull. Soc. R. Bot. Belg.*, vol. xvii, 1879.

I have quoted (p. 71) Dr Alefeld's statement that none of the American species of Linum are heterostyled. This statement was disputed by Kuhn (*Bot. Zeit.*, 1866, p. 201), but has since been confirmed by Dr Ign. / Urban in *Linnaea*, vol. vii, p. 621. Mr Meehan (*Bull. Torrey Bot. Club*, vol. vi, p. 189) has endeavoured to throw doubts on my observations on the sterility of the forms of *L. perenne* when fertilized with their own-form pollen, because a plant from Colorado yielded seed, when growing by itself; but as might have been expected, and as is sufficiently clear from the remarks of a well-known reviewer in the *American Journal of Science*, Mr Meehan mistook *L. Lewisii*, which is not heterostyled, for *L. perenne*.

In the Boragineae, *Lithospermum canescens* differs, according to Mr Erwin F. Smith (*Bot. Gazette*, United States, vol. iv, 1879, p. 168), from the heterostyled species of the same genus by occasionally presenting a mid-styled form, which has a short pistil like that of the short-styled, and short stamens like those of the long-styled form. All the forms seem variable, and the whole case requires further investigation.

Mr Alex. S. Wilson informs me that on comparing the pollen-grains from a long-styled plant of *Erythraea centaurium* with those from some short-styled plants from the island of Arran, they differed in size and shape, as in the case of the undoubtedly heterostyled *Menyanthes trifoliata*, a member of the same family of the Gentianeae. I had myself formerly observed that the flowers on different plants differed much in structure, but could not make out that they presented two distinct forms.

The Rubiaceae contain many more heterostyled plants than any other family, and several additional cases can now be added. Mr C. B. Clarke has been so kind as to send me sketches made in India of two extremely distinct forms of *Adenosacme longifolia*. He remarks 'that the peculiarity of the case is not the difference in the length of the style and stamens / in the two forms, but the extreme difference in the point of insertion of the stamens'. A mid-styled form exists having a short pistil and short stamens seated on the same level, only a little way up the tube of the corolla. Mr Clarke adds that heterostylism is quite common in the Coffee tribe. Mr Hiern, in his observations on the Rubiaceae of tropical Africa (*Journal Linn. Soc. Bot.*, vol. xvi, 1877, p. 252), remarks that dimorphism occurs commonly, or at least in some species, in four or five genera in the tribe of Hedyotideae. Mr M. S. Evans states (*Nature*, 19 September, 1878, p. 543) that in Natal there is a heterostyled Rubiaceous plant, which occasionally, though rarely, presents a third form, and in this the pistil and stamens are of equal length and both are exserted from the mouth of the corolla. He adds that he has found four other heterostyled dimorphic plants, and one of these is a monocotyledon.

Lastly, I have given (p. 97) *Bouvardia leiantha* as doubtfully heterostyled; Mr Bailey has now sent me dried specimens, and as far as the lengths of the pistil and stamens are concerned the species is clearly heterostyled; but no difference could be detected in the size of the pollen-grains; so the case must remain doubtful.

With respect to trimorphic heterostyled plants, Dr Koehne, who has described the Lythraceae of Brazil, has, with great kindness, sent me a long account of them. He knows twenty-one species which are heterostyled, and 340 which are homostyled. He informs me that *Lythrum thymifolia* is not heterostyled, and that I must have received some other species under this name. There are many dimorphic species in America. *Pemphis acidula* is distinctly dimorphic, so are some species of Rotala and Nesaea; thus two / new heterostyled genera are added to the family. Dr Koehne does not believe that any species of Lagerstroemia is, or has been, heterostyled and trimorphic. He has also sent me an outline of an important view, well worth following out, namely, that heterostylism has arisen through the modification of plants which were tending to become polygamous or dioecious.

It is stated at p. 134 that Mr Leggett felt some doubt whether *Pontederia cordata* was trimorphic and heterostyled; but he has since written to me that his doubts are removed: see also to this effect, in *Bull. Torrey Bot. Club*, vol. vi, 1877, p. 170. All three forms of this Pontederia appear to be highly variable. He informs me that humblebees are the fertilizers.

With respect to the origin of the dioecious state, which is discussed in the beginning of the seventh chapter, Hermann Müller has given some interesting remarks in *Kosmos*, 1877, p. 290. The same author shows (ibid., p. 130) that *Valeriana dioica* exists under four forms closely allied to the four presented by Rhamnus, as described in this same chapter. It is much to be desired that someone should experimentise on these forms and make out their meaning. Bernet has published (*Bull. Soc. Bot. France*, vol. xxv, 1878) a paper entitled, 'Disjonction des sexes dans *l'Euonymus Europaeus*', which may be compared with my observations on the same plant. I have stated at p. 215 that I could never find an hermaphrodite plant of the common holly, but according to Mr Hibberd (*Gard. Chron.*, 1877, pp. 39 and 776) such occur among the many cultivated varieties. The evidence, however, is far from conclusive, for it does not appear that Mr Hibberd ever observed under the microscope pollen taken from a plant known to produce berries. Trees of *Juglans cinerea* in the / U. States are monoecious, and like those of *J. regia* consist of two sets, one 'being proterandrous and the other proterogynous (Mr C. G. Pringle, in *Bot. Gazette*, vol. iv, 1879, p. 237); and thus the cross-fertilization of distinct trees is insured. Mr Alex S. Wilson informs me that *Silene inflata* is polygamous on Ben Lawers, as he found hermaphrodite, male and female plants. The case is here mentioned because the flowers on the females are small like those on the females in the gyno-dioecious subclass. In an article in the *Bull. Torrey Bot. Club*, July, 1871, this Silene is, however, said to gyno-dioecious. *Asparagus officinalis* is also polygamous, and the female flowers are about half the size of the male ones; see *Gard. Chron.*, 25 May, 1878; also Breitenbach in *Bot. Zeitung*, 1878, p. 163.

Several cases can now be added to my list of gynodioecious plants, or those which exist as hermaphrodite and female individuals; namely, according to Mr Whitelegge (*Nature*, 3 October, 1878, p. 588), *Stachys germanica, Ranunculus, acris, repens* and *bulbosus*. H. Müller found on the Alps (*Nature*, 1878, p. 516) *Geranium sylvaticum* and *Dianthus superbus* in this state, and the female flowers of the former were of small size. So it is with *Salvia pratensis*, as he informs me in a letter. I have received an additional account of *Plantago lanceolata* being gyno-dioecious in England; and Dr F. Ludwig of Greiz has sent me a description of no less than five forms of this plant which graduate into one another; the intermediate forms being comparatively rare, whilst the hermaphrodite form is the commonest. With respect to the steps by which a gyno-dioecious condition has been gained, H. Müller maintains by many able arguments (*Kosmos*, 1877, pp. 23, 128 and 290) the view which he has propounded; and several botanists think it more probable than the / one advanced by me, see, for instance, *Journal of Botany*, December, 1877, p. 376.

I have stated (p. 9) that after enquiring from several botanists I could hear of no instance, except a doubtful one, of plants in an andro-dioecious condition, or existing as hermaphrodite and male indi-viduals. But H. Müller (*Nature*, 12 September, 1878, p. 159) has found on the Alps *Veratrum album, Dryas octopetala* and *Geum reptans* in this condition. It is an interesting fact that the corollas of the male flowers are not diminished in size like those of the females of gyno-dioecious plants. Asa Gray has also reason to suspect that *Diospyros virginiana* may be andro-dioecious.

The eighth chapter is devoted to cleistogamic flowers, and I have struck out of the list there inserted four genera, owing to information given me by Mr Bentham and Asa Gray. On the other hand, fifteen genera have been added. Mr Bentham informs me that the S. American *Trifolium polymorphum* produces true cleistogamic flowers. Dalibarda, Milium and Vilfa have been added to the list on the authority of A. Gray in a review of this book in the *American Journal of Science*. The cleistogamic flowers of Danthonia are described by Pringle in the *American Naturalist*, 1878, p. 248, and those of another Gramineous genus, Diplachne, by Ascherson in *Sitzungsb. der Gesell. Natur. Freunde, Berlin*, 21 December, 1869. Krascheninikovia has been added from some remarks made in *Journal of Botany*, 1877, p. 377. Batalin has published an essay (*Act. Hort. Petropol*, vol. v, fasc. 2, 1878), 'Kleisto-gamische Blüthen ber Caryophylleen', namely, on Cerastium and Polycarpon. F. Ludwig has described the cleistogamic flowers of *Collomia grandiflora* in *Sitzb. Bot. Vereins. Brandenburg*, 25 August, 1876: see also on same subject Scharlok in *Bot. Zeitung*, 1878, p. 641. A. Grisebach / has discussed fully (*Nachrichten k. Gesell. der Wissen. zu Gottingen*, 1 June, 1878) the cleistogamic flowers produced by *Carda-mine chenopodifolia*, which bury themselves in the ground. See also on same subject Drude in *Sitzb. der Versamml. d. Naturf. in Cassel*, 1878. From a note received from Dr Koehne it is clear that *Ammannia latifolia* bears cleistogamic flowers. According to Mr Bessey (*The American Naturalist*, 1878, p. 69) this is likewise the case with *Lithospermum longiflorum*. Three genera of Orchideae have been added to the list, from information given me by Mr Spencer Moore and from some remarks in *Journal of Botany*, 1877, p. 377. Lastly, Mr Bennett has published (*Journal Linn. Soc. Bot.*, no. 101, 1879) some additional 'notes on cleistogamic flowers', chiefly on those of Viola and Impatiens.

With respect to the statement (p. 237) on the authority of Mr Wallis, that *Drosera rotundifolia* opens its flowers only early in the morning, Mr Conybeare informs me that he once saw in Cornwall, at 2 p.m., the ground 'starred over with the fully-expanded flowers of this plant'. He had previously long endeavoured to find a plant with open flowers.

The number of species in which pods produced by cleistogamic flowers bury themselves in the ground is remarkable. I have attributed (p. 243) this action to the advantage gained by their protection from various enemies, and much may be said in favour of this view; but Mr W. Thiselton Dyer, in an interesting article (*Nature*, 4 April, 1878, p. 446) has called attention to some observations made long ago by Mr Bentham (*Catalogue des Plantes indig. des Pyrénées*, 1826, p. 85) on the fruiting of *Helianthemum prostratum*. He believes, as does Mr Dyer, that the capsules of this helianthemum and some other plants (for instance, of / cyclamen) are kept cool and moist by being laid on the ground; they thus mature more slowly and are enabled to grow to a larger size. In this simple action we probably see the first step to the further development of the process, and to the capsules burying themselves beneath the surface. In some cases the difference between the subaerial and subterranean pods on the same plant and both produced by cleistogamic flowers is extraordinary: Mr Meehan sent me three subterranean pods of *Amphicarpaea monoica*, each containing a single large seed; and my own plants produced several subaerial pods, each containing from one to three small seeds. These latter weighed on an average only $\frac{1}{70}$ of the subterranean seeds! This difference, however, is not quite accurate, as the coats of the subterranean pods adhered so firmly to the seeds that they were not removed and were weighed with them; but from their thinness and lightness they could not have much affected the result. /

CONTENTS

CHAPTER I

Heterostyled dimorphic plants: Primulaceae

CHAPTER II

Hybrid primulas

CHAPTER III

Heterostyled dimorphic plants – continued

CHAPTER IV

Heterostyled trimorphic plants

CHAPTER V

Illegitimate offspring of heterostyled plants

CHAPTER VI

Concluding remarks on heterostyled plants

CHAPTER VII

Polygamous, dioecious, and gyno-dioecious plants

CHAPTER VIII

Cleistogamic flowers

INTRODUCTION

The subject of the present volume, namely the differently formed flowers normally produced by certain kinds of plants, either on the same stock or on distinct stocks, ought to have been treated by a professed botanist, to which distinction I can lay no claim. As far as the sexual relations of flowers are concerned, Linnaeus long ago divided them into hermaphrodite, monoecious, dioecious, and polygamous species. This fundamental distinction, with the aid of several subdivisions in each of the four classes, will serve my purpose; but the classification is artificial, and the groups often pass into one another.

The hermaphrodite class contains two interesting sub groups, namely, heterostyled and cleistogamic plants; but there are several other less important subdivisions, presently to be given, in which flowers differing in various ways from one another are produced by the same species.

Some plants were described by me several years ago, in a series of papers read before the Linnean Society,[1] / the individuals of which exist under two or three forms, differing in the length of their pistils and stamens and in other respects. They are called by me dimorphic and trimorphic, but have since been better named by Hildebrand, heterostyled.[2] As I have many still unpublished observations with respect to

---

[1] 'On the Two Forms, or Dimorphic Condition in the Species of Primula, and on their remarkable Sexual Relations.' *Journal of the Proceedings of the Linnean Society*, vol. vi, 1862, p. 77.

'On the Existence of Two Forms, and on their Reciprocal Sexual Relation, in several Species of the Genus Linum.' Ibid., vol. vii, 1863, p. 69.

'On the Sexual Relations of the Three Forms of *Lythrum salicaria*.' Ibid., vol. viii, 1864, p. 169.

'On the Character and Hybrid-like Nature of the Offspring from the Illegitimate Unions of Dimorphic and Trimorphic Plants.' Ibid., vol. x, 1868, p. 393.

'On the Specific Differences between *Primula veris*, Brit. Fl. (var. *officinalis*, Linn.), *P. vulgaris*, Brit. Fl. (var. *acaulis*, Linn.), and *P. elatior*, Jacq.; and on the Hybrid Nature of the Common Oxlip. With Supplementary Remarks on Naturally Produced Hybrids in the Genus Verbascum.' Ibid., vol. x, 1868, p. 437.

[2] The term 'heterostyled' does not express all the differences between the forms; but this is a failure common in many cases. As the term has been adopted by writers in

1

these plants, it has seemed to me advisable to republish my former papers in a connected and corrected form, together with the new matter. It will be shown that these heterostyled plants are adapted for reciprocal fertilization; so that the two or three forms, though all are hermaphrodites, are related to one another almost like the males and females of ordinary unisexual animals. I will also give a full abstract of such observations as have been published since the appearance of my papers; but only those cases will be noticed, with respect to which the evidence seems fairly satisfactory. Some plants have been supposed to be heterostyled merely from their pistils and stamens varying greatly in length, and I have been myself more than once thus deceived. With some species the / pistil continues growing for a long time, so that if old and young flowers are compared they might be thought to be heterostyled. Again, a species tending to become dioecious, with the stamens reduced in some individuals and with the pistils in others, often presents a deceptive appearance. Unless it be proved that one form is fully fertile only when it is fertilized with pollen from another form, we have not complete evidence that the species is heterostyled. But when the pistils and stamens differ in length in two or three sets of individuals, and this is accompanied by a difference in the size of the pollen grains or in the state of the stigma, we may infer with much safety that the species is heterostyled. I have, however, occasionally trusted to a difference between the two forms in the length of the pistil alone, or in the length of the stigma together with its more or less papillose condition; and in one instance differences of this kind have been proved by trials made on the fertility of the two forms, to be sufficient evidence.

The second sub group above referred to consists of hermaphrodite plants, which bear two kinds of flowers – the one perfect and fully expanded – the other minute, completely closed, with the petals rudimentary, often with some of the anthers aborted, and the remaining ones together with the stigmas much reduced in size; yet these flowers are perfectly fertile. They have been called by Dr Kuhn[3] cleistogamic,

---

various countries, I am unwilling to change it for that of *heterogone* or *heterogonous*, though this has been proposed by so high an authority as Professor Asa Gray: see the *American Naturalist*, January, 1877, p. 42.

[3] *Botanische Zeitung*, 1867, p. 65. Several plants are known occasionally to produce flowers destitute of a corolla; but they belong to a different class of cases from cleistogamic flowers. This deficiency seems to result from the conditions to which the plants have been subjected, and partakes of the nature of a monstrosity. All the flowers on the same plant are commonly affected in the same manner. Such cases, though they have sometimes been ranked as cleistogamic, do not come within our present scope: see Dr Maxwell Masters, *Vegetable Teratology*, 1869, p. 403.

and they / will be described in the last chapter of this volume. They are manifestly adapted for self-fertilization, which is effected at the cost of a wonderfully small expenditure of pollen; whilst the perfect flowers produced by the same plant are capable of cross-fertilization. Certain aquatic species, when they flower beneath the water, keep their corollas closed, apparently to protect their pollen; they might there-fore be called cleistogamic, but for reasons assigned in the proper place are not included in the present sub group. Several cleistogamic species, as we shall hereafter see, bury their ovaries or young capsules in the ground. Some few plants produce subterranean flowers, as well as ordinary ones; and these might have been formed into a small separate subdivision.

Another interesting subdivision consists of certain plants, discovered by H. Müller, some individuals of which bear conspicuous flowers adapted for cross-fertilization by the aid of insects, and others much smaller and less conspicuous flowers, which have often been slightly modified so as to ensure self-fertilization. *Lysimachia vulgaris*, *Euphrasia officinalis*, *Rhinanthus crista-galli*, and *Viola tricolor* come under this head.[4] The smaller and less conspicuous flowers are not closed, but as far as the purpose which they serve is concerned, namely, the assured propagation of the species, they approach in nature cleistogamic flowers; but they differ from them by the two kinds being produced on distinct plants.

With many plants, the flowers towards the outside of the infloresc-ence are much larger and more conspicuous than the central ones. As I shall not have occasion / to refer to plants of this kind in the following chapters, I will here give a few details respecting them. It is familiar to every one that the ray-florets of the Compositae often differ remark-ably from the others; and so it is with the outer flowers of many Umbelliferae, some Cruciferae and a few other families. Several species of Hydrangea and Viburnum offer striking instances of the same fact. The Rubiaceous genus Mussaenda, presents a very curious appearance from some of the flowers having the tip of one of the sepals developed into a large petal-like expansion, coloured either white or purple. The outer flowers in several Acantaceous genera are large and conspicuous, but sterile; the next in order are smaller, open, moderately fertile and capable of cross-fertilization; whilst the central

[4] H. Müller, *Nature*, 25 September, 1873 (vol. viii), p. 433, and 20 November, 1873 (vol. ix), p. 44. Also *Die Befruchtung der Blumen*, etc., 1873, p. 294.

ones are cleistogamic, being still smaller, closed and highly fertile; so that here the inflorescence consists of three kinds of flowers.[5] From what we know in other cases of the use of the corolla, coloured bracteae, etc., and from what H. Müller has observed[6] on the frequency of the visits of insects to the flower-heads of the Umbelliferae and Compositae being largely determined by their conspicuousness, there can be no doubt that the increased size of the corolla of the outer flowers, the inner ones being in all the above cases small, serves to attract insects. The result is that cross-fertilization is thus favoured. Most flowers wither soon after being fertilized, but Hildebrand states[7] that the ray-florets of the Compositae last for a long time, until all those on the disc are impregnated; and this clearly shows the use of the former. The ray-florets, / however, are of service in another and very different manner, namely, by folding inwards at night and during cold rainy weather, so as to protect the florets of the disc.[8] Moreover they often contain matter which is excessively poisonous to insects, as may be seen in the use of flea-powder, and in the case of Pyrethrum. M. Belhomme has shown that the ray-florets are more poisonous than the disc-florets in the ratio of about three to two. We may therefore believe that the ray-florets are useful in protecting the flowers from being gnawed by insects.[9]

It is a well-known yet remarkable fact that the circumferential flowers of many of the foregoing plants have both their male and female reproductive organs aborted, as with the Hydrangea, Viburnum and certain Compositae; or the male organs alone are aborted, as in many Compositae. Between the sexless, female, and hermaphrodite states of these latter flowers, the finest gradations may be traced, as Hildebrand has shown.[10] He also shows that there is a close relation

---

[5] J. Scott, *Journal of Botany*, London, new series, vol. i, 1872, pp. 161–4.

[6] *Die Befruchtung, der Blumen*, pp. 108, 412.

[7] See his interesting memoir, *Ueber die Geschlechtsverhältnisse bei den Compositen*, 1869, p. 92.

[8] Kerner clearly shows that this is the case: *Die Schutzmittel des Pollens*, 1873, p. 28.

[9] *Gardener's Chronicle*, 1861, p. 1067. Lindley, *Vegetable Kingdom*, on Chrysanthemum, 1853, p. 706. Kerner in his interesting essay (*Die Schutzmittel der Blüthen gegen unberufene Gäste*, 1875, p. 19) insists that the petals of most plants contain matter which is offensive to insects, so that they are seldom gnawed, and thus the organs of fructification are protected. My grandfather in 1790 (*Loves of the Plants*, canto iii, note to lines 184, 188) remarks that 'The flowers or petals of plants are perhaps in general more acrid than their leaves; hence they are much seldomer eaten by insects.'

[10] *Ueber die Geschlechtsverhältnisse bei den Compositen*, 1869, pp. 78–91.

between the size of the corolla in the ray-florets and the degree of abortion in their reproductive organs. As we have good reason to believe that these florets are highly serviceable to the plants which possess them, more especially by rendering the flower-heads conspicuous / to insects, it is a natural inference that their corollas have been increased in size for this special purpose; and that their development has subsequently led, through the principle of compensation or balancement, to the more or less complete reduction of the reproductive organs. But an opposite view may be maintained, namely, that the reproductive organs first began to fail, as often happens under cultivation,[11] and, as a consequence, the corolla became, through compensation, more highly developed. This view, however, is not probable, for when hermaphrodite plants become dioecious or gyno-dioecious – that is, are converted into hermaphrodites and females – the corolla of the female seems to be almost invariably reduced in size in consequence of the abortion of the male organs. The difference in the result in these two classes of cases may perhaps be accounted for by the matter saved through the abortion of the male organs in the females of gyno-dioecious and dioecious plants being directed (as we shall see in a future chapter) to the formation of an increased supply of seeds; whilst in the case of the exterior florets and flowers of the plants which we are here considering, such matter is expended in the development of a conspicuous corolla. Whether in the present class of cases the corolla was first affected, as seems to me the more probable view, or the reproductive organs first failed, their states of development are now firmly correlated. We see this well illustrated in Hydrangea and Viburnum; for when these plants are cultivated, the corollas of both the interior and exterior flowers become largely developed, and their reproductive organs are aborted. /

There is a closely analogous subdivision of plants, including the genus Muscari (or Feather Hyacinth) and the allied Bellevalia, which bear both perfect flowers and closed bud-like bodies that never expand. The latter resemble in this respect cleistogamic flowers, but differ widely from them in being sterile and conspicuous. Not only the aborted flower-buds and their peduncles (which are elongated apparently through the principle of compensation) are brightly coloured, but so is the upper part of the spike – all, no doubt, for the sake of

[11] I have discussed this subject in my *Variation of Animals and Plants under Domestication*, chap. xviii, 2nd edit., vol. ii, pp. 152, 156.

guiding insects to the inconspicuous perfect flowers. From such cases as these we may pass on to certain Labiatae, for instance, *Salvia Horminum*, in which (as I hear from Mr Thiselton Dyer) the upper bracts are enlarged and brightly coloured, no doubt for the same purpose as before, with the flowers suppressed.

In the carrot and some allied Umbelliferae, the central flower has its petals somewhat enlarged, and these are of a dark purplish-red tint; but it cannot be supposed that this one small flower makes the large white umbel at all more conspicuous to insects. The central flowers are said[12] to be neuter or sterile, but I obtained by artificial fertilization a seed (fruit) apparently perfect from one such flower. Occasionally two or three of the flowers next to the central one are similarly characterized; and according to Vaucher[13] 'cette singulière dégénération s'étend quelquefois á l'ombelle entière'. That the modified central flower is of no functional importance to the plant is almost certain. It may perhaps be a remnant of a former and ancient condition of the species, when one flower alone, the / central one, was female and yielded seeds, as in the umbelliferous genus Echinophora. There is nothing surprising in the central flower tending to retain its former condition longer than the others; for when irregular flowers become regular or peloric, they are apt to be central; and such peloric flowers apparently owe their origin either to arrested development – that is, to the preservation of an early stage of development – or to reversion. Central and perfectly developed flowers in not a few plants in their normal condition (for instance, the common rue and adoxa) differ slightly in structure, as in the number of the parts, from the other flowers on the same plant. All such cases seem connected with the fact of the bud which stands at the end of the shoot being better nourished than the others, as it receives the most sap.[14]

The cases hitherto mentioned relate to hermaphrodite species which bear differently constructed flowers; but there are some plants that produce differently formed seeds, of which Dr Kuhn has given a list.[15] With the Umbelliferae and Compositae, the flowers that produce these seeds likewise differ, and the differences in the structure of the seeds

---

[12] *The English Flora*, by Sir J. E. Smith, 1824, vol. ii, p. 39.

[13] *Hist. Phys. des Plantes d'Europe*, 1841, vol. ii, p. 614. On the Echinophora, p. 627.

[14] This whole subject, including pelorism, has been discussed, and references given, in my *Variation of Animals and Plants under Domestication*, chap. xxvi, 2nd edit., vol. ii, p. 338.

[15] *Bot. Zeitung*, 1867, p. 67.

are of a very important nature. The causes which have led to differences in the seeds on the same plant are not known; and it is very doubtful whether they subserve any special end.

We now come to our second Class, that of monoecious species, or those which have their sexes separated but borne on the same plant. The flowers necessarily differ, but when those of one sex include rudiments / of the other sex, the difference between the two kinds is usually not great. When the difference is great, as we see in catkin-bearing plants, this depends largely on many of the species in this, as well as in the next or dioecious class, being fertilized by the aid of the wind;[16] for the male flowers have in this case to produce a surprising amount of incoherent pollen. Some few monoecious plants consist of two bodies of individuals, with their flowers differing in function, though not in structure; for certain individuals mature their pollen before the female flowers on the same plant are ready for fertilization, and are called proterandrous; whilst conversely other individuals, called pro terogynous, have their stigmas mature before their pollen is ready. The purpose of this curious functional difference obviously is to favour the cross-fertilization of distinct plants. A case of this kind was first observed by Delpino in the walnut (*Juglans regia*), and has since been observed with the common nut (*Corylus avellana*). According to H. Müller the individual plants of a few hermaphrodite species differ in a like manner; some being proterandrous and others proterogynous.[17] On cultivated trees of the walnut and mulberry, the male flowers have been observed to abort on certain individuals,[18] which have thus been converted into females; but whether there are any species in a state of nature which co-exist as monoecious and female individuals, I do not know.

The third Class consists of dioecious species, and the / remarks made under the last class with respect to the amount of difference between the male and female flowers are here applicable. It is at present an inexplicable fact that with some dioecious plants, of which the

[16] Delpino, *Studi sopra uno Lignaggio Anemofilo*, Firenze, 1871.
[17] Delpino, *Ult. Osservazioni sulla Dicogamia*, part ii, fasc. ii, p. 337. Mr Wetterhan and H. Müller on Corylus, *Nature*, vol. xi, p. 507, and 1875, p. 26. On proterandrous and proterogynous hermaphrodite individuals of the same species, see H. Müller, *Die Befruchtung*, etc., pp. 285, 339.
[18] *Gardener's Chron.*, 1847, pp. 541, 558.

7

Restiaceae of Australia and the Cape of Good Hope offer the most striking instance, the differentiation of the sexes has affected the whole plant to such an extent (as I hear from Mr Thiselton Dyer) that Mr Bentham and Professor Oliver have often found it impossible to match the male and female specimens of the same species. In my seventh chapter some observations will be given on the gradual conversion of heterostyled and of ordinary hermaphrodite plants into dioecious or sub-dioecious species.

The fourth and last Class consists of the plants which were called polygamous by Linnaeus; but it appears to me that it would be convenient to confine this term to the species which co-exist as hermaphrodites, males, and females; and to give new names to several other combinations of the sexes – a plan which I shall here follow. Polygamous plants, in this confined sense of the term, may be divided into two sub groups, according as the three sexual forms are found on the same individual or on distinct individuals. Of this latter or trioicous sub group, the common ash (*Fraxinus excelsior*) offers a good instance: thus, I examined during the spring and autumn fifteen trees growing in the same field; and of these, eight produced male flowers alone, and in the autumn not a single seed; four produced only female flowers, which set an abundance of seeds; three were hermaphrodites, which had a different aspect from the other trees whilst in flower, and two of them produced nearly as many seeds as the female trees, whilst the third produced none, so that it / was in function a male. The separation of the sexes, however, is not complete in the ash; for the female flowers include stamens, which drop off at an early period, and their anthers, which never open or dehisce, generally contain pulpy matter instead of pollen. On some female trees, however, I found a few anthers containing pollen grains apparently sound. On the male trees most of the flowers include pistils, but these likewise drop off at an early period; and the ovules, which ultimately abort, are very small compared with those in female flowers of the same age.

Of the other or monoecious sub group of polygamous plants, or those which bear hermaphrodite, male and female flowers on the same individual, the common maple (*Acer campestre*) offers a good instance; but Lecoq states[19] that some trees are truly dioecious, and this shows how easily one state passes into another.

[19] *Géographie Botanique*, vol. v, p. 367.

A considerable number of plants generally ranked as polygamous exist under only two forms, namely, as hermaphrodites and females; and these may be called gyno-dioecious, of which the common thyme offers a good example. In my seventh chapter I shall give some observations on plants of this nature. Other species, for instance several kinds of atriplex, bear on the same plant hermaphrodite and female flowers; and these might be called gyno-monoecious, if a name were desirable for them.

Again there are plants which produce hermaphrodite and male flowers on the same individual, for instance, some species of Galium, Veratrum, etc.; and these might be called andro-monoecious. If there exist plants, the individuals of which consist of hermaphrodites and males, these might be distinguished / as andro-dioecious. But, after making enquiries from several botanists, I can hear of no such cases. Lecoq, however, states,[20] but without entering into full details, that some plants of *Caltha palustris* produce only male flowers, and that these live mingled with the hermaphrodites. The rarity of such cases as this last one is remarkable, as the presence of hermaphrodite and male flowers on the same individual is not an unusual occurrence; it would appear as if nature did not think it worthwhile to devote a distinct individual to the production of pollen, excepting when this was indispensably necessary, as in the case of dioecious species.

I have now finished my brief sketch of the several cases, as far as known to me, in which flowers differing in structure or in function are produced by the same species of plant. Full details will be given in the following chapters with respect to many of these plants. I will begin with the heterostyled, then pass on to certain dioecious, sub-dioecious, and polygamous species, and end with the cleistogamic. For the convenience of the reader, and to save space, the less important cases and details have been printed in smaller type.

I cannot close this Introduction without expressing my warm thanks to Dr Hooker for supplying me with specimens and for other aid; and to Mr Thiselton Dyer and Professor Oliver for giving me much information and other assistance. Professor Asa Gray, also, has uniformly aided me in many ways. To Fritz Müller of St Catharina, in Brazil, I am indebted for many dried flowers of heterostyled plants, often accompanied with valuable notes. /

[20] *Géographie Botanique*, vol. iv, p. 488.

CHAPTER I

HETEROSTYLED DIMORPHIC PLANTS:
PRIMULACEAE

Primula veris or the cowslip – Differences in structure between the two forms – Their degrees of fertility when legitimately and illegitimately united – *P. elatior, vulgaris, Sinensis, auricula*, etc. – Summary on the fertility of the heterostyled species of primula – Homostyled species of primula – *Hottonia palustris* – *Androsace Vitalliana*.

It has long been known to botanists that the common cowslip (*Primula veris*, Brit. Flora, var. *officinalis*, Lin.) exists under two forms, about equally numerous, which obviously differ from each other in the length of their pistils and stamens.[1] This difference has hitherto been looked at as a case of mere variability, but this view, as we shall presently see, is far from the true one. Florists who cultivate the polyanthus and auricula have long been aware of the two kinds of flowers, and they call the plants which display the globular stigma at the mouth of the corolla, 'pin-headed' or 'pin-eyed', and those which display the anthers, 'thrum-eyed'.[2] I will designate the two forms as the long-styled and short-styled.

The pistil in the long-styled form is almost exactly twice as long as that of the short-styled. The stigma / stands in the mouth of the corolla, or projects just above it, and is thus externally visible. It stands high above the anthers, which are situated halfway down the tube and cannot be easily seen. In the short-styled form the anthers are attached near the mouth of the tube and therefore stand above the stigma, which is seated in about the middle of the tubular corolla. The corolla itself is of a different shape in the two forms; the throat or expanded

---

[1] This fact, according to von Mohl (*Bot. Zeitung*, 1863, p. 326) was first observed by Persoon in the year 1794.

[2] In Johnson's *Dictionary*, *thrum* is said to be the ends of weavers' threads; and I suppose that some weaver who cultivated the polyanthus invented this name, from being struck with some degree of resemblance between the cluster of anthers in the mouth of the corolla and the ends of his threads.

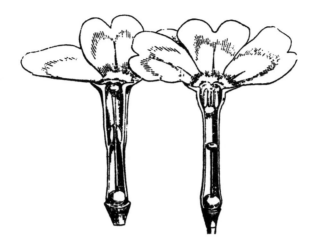

Long-styled form   Short-styled form

Fig. 1
Primula veris

portion above the attachment of the anthers being much longer in the long-styled than in the short-styled form. Village children notice this difference, as they can best make necklaces by threading and slipping the corollas of the long-styled flowers into one another. But there are much more important differences. The stigma in the long-styled form / is globular; in the short-styled it is depressed on the summit, so that the longitudinal axis of the former is sometimes nearly double that of the latter. Although it is somewhat variable in shape, one difference is persistent, namely, roughness: in some specimens carefully compared, the papillae which render the stigma rough were in the long-styled form from twice to thrice as long as in the short-styled. The anthers do not differ in size in the two forms, which I mention, because this is the case with some heterostyled plants. The most remarkable difference is in the pollen grains. I measured with the micrometer many specimens, both dry and wet, taken from plants growing in different situations, and always found a palpable difference. The grains distended with water from the short-styled flowers were about 0·038 mm ($^{10-11}$/$_{7000}$ of an inch) in diameter, whilst those from the long-styled were about 0·0254 mm (^{7}/$_{1000}$ of an inch), which is in the ratio of 100 to 67. The pollen grains, therefore, from the longer stamens of the short-styled form are plainly larger than those from the shorter stamens of the long-styled. When examined dry, the smaller grains are seen under a low power to be more transparent than the larger grains, and apparently in a greater degree than can be accounted for by their less diameter. There is also a difference in shape, the grains from the short-styled plants being nearly spherical, those from the long-styled being oblong with the angles rounded; this difference disappears when the grains are distended with water. The long-styled plants generally tend to flower a little before the short-styled: for instance I had twelve plants of each form growing in separate pots and treated in every respect alike; and at the time when only a single short-styled plant was in / flower, seven of the long-styled had expanded their flowers.

We shall, also, presently see that the short-styled plants produce more seed than the long-styled. It is remarkable, according to Professor Oliver,[3] that the ovules in the unexpanded and unimpregnated flowers of the latter are considerably larger than those of the short-styled flowers; and this I suppose is connected with the long-styled flowers producing fewer seeds, so that the ovules have more space and nourishment for rapid development.

To sum up the difference: The long-styled plants have a much longer pistil, with a globular and much rougher stigma, standing high above the anthers. The stamens are short; the grains of pollen smaller and oblong in shape. The upper half of the tube of the corolla is more expanded. The number of seeds produced is smaller and the ovules larger. The plants tend to flower first.

The short-styled plants have a short pistil, half the length of the tube of the corolla, with a smooth depressed stigma standing beneath the anthers. The stamens are long; the grains of pollen are spherical and larger. The tube of the corolla is of uniform diameter except close to the upper end. The number of seeds produced is larger.

I have examined a large number of flowers; and though the shape of the stigma and the length of the pistil both vary, especially in the short-styled form, I have never met with any transitional states between the two forms in plants growing in a state of nature. There is never the slightest doubt under which form a plant ought to be classed. The two kinds of flowers are / never found on the same individual plant. I marked many cowslips and primroses, and on the following year all retained the same character, as did some in my garden which flowered

[3] *Nat. Hist. Review*, July, 1862, p. 237.

out of their proper season in the autumn. Mr W. Wooler, of Darling-
ton, however, informs us that he has seen early blossoms on the polyanthus,[4] which were not long-styled, but became so later in the season.
Possibly in this case the pistils may not have been fully developed
during the early spring. An excellent proof of the permanence of the
two forms may be seen in nursery-gardens, where choice varieties of
the polyanthus are propagated by division; and I found whole beds of
several varieties, each consisting exclusively of the one or the other
form. The two forms exist in the wild state in about equal numbers: I
collected 522 umbels from plants growing in several stations, taking a
single umbel from each plant; and 241 were long-styled, and 281
short-styled. No difference in tint or size could be perceived in the two
great masses of flowers.

We shall presently see that most of the species of primula exist
under two analogous forms; and it may be asked what is the meaning
of the above-described important differences in their structure? The
question seems well worthy of careful investigation, and I will give my
observations on the cowslip in detail. The first idea which naturally
occurred to me was, that this species was tending towards a dioecious
condition; that the long-styled plants, with their longer pistils, rougher
stigmas, and smaller pollen grains, were more feminine in nature, and
would produce more seed; that the short-styled plants, with their
shorter pistils, longer stamens and larger pollen / grains, were more
masculine in nature. Accordingly, in 1860, I marked a few cowslips of
both forms growing in my garden, and others growing in an open
field, and others in a shady wood, and gathered and weighed the seed.
In all the lots the short-styled plants yielded, contrary to my expecta-
tion, most seed. Taking the lots together, the following is the result:

TABLE 1

	Number of plants	Number of umbels produced	Number of capsules produced	Weight of seed in grains
Short-styled cowslips	9	33	199	83
Long-styled cowslips	13	51	261	91

If we compare the weight from an equal number of plants, and from

[4] I have proved by numerous experiments, hereafter to be given, that the
polyanthus is a variety of *Primula veris*.

13

an equal number of umbels, and from an equal number of capsules of the two forms, we get the following results:

TABLE 2

	Number of plants	Weight of seed in grains	Number of umbels	Weight of seed	Number of capsules	Weight of seed in grains
Short-styled cowslips	10	92	100	251	100	41
Long-styled cowslips	10	70	100	178	100	34

So that, by all these standards of comparison, the short-styled form is the more fertile; if we take the number of umbels (which is the fairest standard, for large and small plants are thus equalized), the short-styled plants produce more seed than the long-styled, in the proportion of nearly four to three.

In 1861 the trial was made in a fuller and fairer / manner. A number of wild plants had been transplanted during the previous autumn into a large bed in my garden, and all were treated alike; the result was:

TABLE 3

	Number of plants	Number of umbels	Weight of seed in grains
Short-styled cowslips	47	173	745
Long-styled cowslips	58	208	692

These figures give us the following proportions:

TABLE 4

	Number of plants	Weight of seed in grains	Number of umbels	Weight of seed in grains
Short-styled cowslips	100	1,585	100	430
Long-styled cowslips	100	1,093	100	332

The season was much more favourable this year than the last; the plants also now grew in good soil, instead of in a shady wood, or struggling with other plants in the open field; consequently, the actual produce of seed was considerably larger. Nevertheless we have the same relative result; for the short-styled plants produced more seed than the long-styled in nearly the proportion of three to two; but if we take the fairest standard of comparison, namely, the product of seeds from an equal number of umbels, the excess is, as in the former case, nearly as four to three.

Looking to these trials made during two successive years on a large number of plants, we may safely conclude that the short-styled form is more productive than the long-styled form, and the same result holds / good with some other species of primula. Consequently my anticipation that the plants with longer pistils, rougher stigmas, shorter stamens, and smaller pollen grains, would prove to be more feminine in nature, is exactly the reverse of the truth.

In 1860 a few umbels on some plants of both the long-styled and short-styled form, which had been covered by a net, did not produce any seed, though other umbels on the same plants, artificially fertilized, produced an abundance of seed; and this fact shows that the mere covering in itself was not injurious. Accordingly, in 1861, several plants were similarly covered just before they expanded their flowers; these turned out as follows:

TABLE 5

—	Number of plants	Number of umbels produced	Product of seed
Short-styled	6	24	1·3 grain weight of seed, or about 50 in number
Long-styled	18	74	Not one seed

Judging from the exposed plants which grew all round in the same bed, and had been treated in the same manner, excepting that they had been exposed to the visits of insects, the above six short-styled plants ought to have produced 92 grains' weight of seed instead of only 1·3; and the eighteen long-styled plants, which produced not one seed, ought to have produced above 200 grains' weight. The production of a few seeds by the short-styled plants was probably due to the action of thrips or of some other minute insect. It is scarcely necessary to give any additional evidence, but I may add that ten pots of polyanthuses and / cowslips of both forms, protected from insects in my greenhouse, did not set one pod, though artificially fertilized flowers in other pots produced an abundance. We thus see that the visits of insects are absolutely necessary for the fertilization of *Primula veris*. If the corolla of the long-styled form had dropped off, instead of remaining attached in a withered state to the ovarium, the anthers attached to the lower part of the tube with some pollen still adhering to them would have been dragged over the stigma, and the flowers would have been partially self-fertilized, as is the case with *Primula Sinensis* through this means. It is a rather curious fact that so trifling a difference as the falling-off of the withered corolla, should make a very great difference in the number of seeds produced by a plant, if its flowers are not visited by insects.

The flowers of the cowslip and of the other species of the genus secrete plenty of nectar; and I have often seen humble-bees, especially *B. hortorum* and *muscorum*, sucking the former in a proper manner,[5] though they sometimes bite holes through the corolla. No doubt moths likewise visit the flowers, as one of my sons caught *Cucullia verbasci* in the act. The pollen readily adheres to any thin object which is inserted into a flower. The anthers in the one form stand nearly, but not exactly, on a level with the stigma of the other; for the distance between the anthers and stigma in the short-styled form is greater than that in the long-styled, in the ratio of 100 to 90. This difference is the result of the anthers in the long-styled form standing rather higher in the tube than does the stigma in the short-styled, and this favours their / pollen being deposited on it. It follows from the position of the organs that if the proboscis of a dead humble-bee, or a thick bristle or rough needle, be pushed down the corolla, first of one form and then of the other, as an insect would do in visiting the two forms growing mingled together, pollen from the long-stamened form adheres round the base of the object, and is left with certainty on the stigma of the long-styled form; whilst pollen from the short stamens of the long-styled form adheres a little way above the extremity of the object, and some is generally left on the stigma of the other form. In accordance with this observation I found that the two kinds of pollen, which could easily be recognized under the microscope, adhered in this manner to the proboscides of the two species of humble-bees and of the moth, which were caught visiting the flowers; but some small grains were mingled with the larger grains round the base of the proboscis, and conversely some large grains with the small grains near the extremity

---

[5] H. Müller has also seen *Anthophora pilipes* and a Bombylius sucking the flowers. *Nature*, 10 December, 1874, p. 111.

of the proboscis. Thus pollen will be regularly carried from the one form to the other, and they will reciprocally fertilize one another. Nevertheless an insect in withdrawing its proboscis from the corolla of the long-styled form cannot fail occasionally to leave pollen from the same flower on the stigma; and in this case there might be self-fertilization. But this will be much more likely to occur with the short-styled form; for when I inserted a bristle, or other such objects in the corolla of this form, and had, therefore, to pass it down between the anthers seated round the mouth of the corolla, some pollen was almost invariably carried down and left on the stigma. Minute insects, such as thrips, which sometimes haunt the flowers, would / likewise be apt to cause the self-fertilization of both forms.

The several foregoing facts led me to try the effects of the two kinds of pollen on the stigmas of the two forms. Four essentially different unions are possible; namely, the fertilization of the stigma of the long-styled form by its own-form pollen, and by that of the short-styled; and the stigma of the short-styled form by its own-form pollen, and by that of the long-styled. The fertilization of either form with pollen from the other form may be conveniently called a *legitimate union*, from reasons hereafter to be made clear; and that of either form with its own-form pollen an *illegitimate union*. I formerly applied the term 'heteromorphic' to the legitimate unions, and 'homomorphic' to the illegitimate unions; but after discovering the existence of trimorphic plants, in which many more unions are possible, these two terms ceased to be applicable. The illegitimate union of both forms might have been tried in three ways; for a flower of either form may be fertilized with pollen from the same flower, or with that from another flower on the same plant, or with that from a distinct plant of the same form. But to make my experiments perfectly fair, and to avoid any evil result from self-fertilization or too close interbreeding, I have invariably employed pollen from a distinct plant of the same form for the illegitimate unions of all the species; and therefore it may be observed that I have used the term 'own-form pollen' in speaking of such unions. The several plants in all my experiments were treated in exactly the same manner, and were carefully protected by fine nets from the access of insects, excepting thrips, which it is impossible to exclude. I performed all the manipulations myself, and weighed the seeds in a chemical balance; but during / many subsequent trials I followed the more accurate plan of counting the seeds. Some of the capsules contained no seeds, or only two or three, and these are excluded in the column headed 'good capsules' in several of the following tables:

TABLE 6   *Primula veris*

Nature of the union	Number of flowers fertilized	Total number of capsules produced	Number of good capsules	Weight of seed in grains	Calculated weight of seed from 100 good capsules
Long-styled by pollen of short-styled.					
Legitimate union	22	15	14	8·8	62
Long-styled by own-form pollen.					
Illegitimate union	20	8	5	2·1	42
Short-styled by pollen of long-styled.					
Legitimate union	13	12	11	4·9	44
Short-styled by own-form pollen.					
Illegitimate union	15	8	6	1·8	30
SUMMARY					
The two legitimate unions	35	27	25	13·7	54
The two illegitimate unions	35	16	11	3·9	35

The results may be given in another form (Table 7) by comparing, first, the number of capsules, whether good or bad, or of the good alone, produced by 100 flowers of both forms when legitimately and illegitimately fertilized; secondly, by comparing the weight of seed in 100 of these capsules, whether good or bad; or, thirdly, in 100 of the good capsules. /

TABLE 7

Nature of the union	Number of flowers fertilized	Number of capsules	Number of good capsules	Weight of seed in grains	Number of capsules	Weight of seed in grains	Number of good capsules	Weight of seed in grains
The two legitimate unions	100	77	71	39	100	50	100	54
The two illegitimate unions	100	45	31	11	100	24	100	35

We here see that the long-styled flowers fertilized with pollen from

the short-styled yield more capsules, especially good ones (i.e. containing more than one or two seeds), and that these capsules contain a
greater proportional weight of seeds than do the flowers of the long-
styled when fertilized with pollen from a distinct plant of the same
form. So it is with the short-styled flowers, if treated in an analogous
manner. Therefore I have called the former method of fertilization a
legitimate union, and the latter, as it fails to yield the full complement
of capsules and seeds, an illegitimate union. These two kinds of union
are graphically represented in Fig. 2.

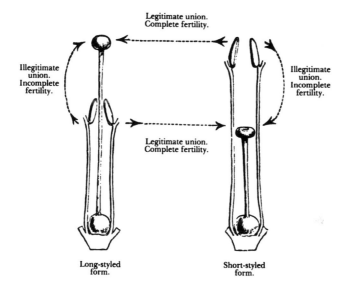

Fig. 2

    If we consider the results of the two legitimate unions taken together
and the two illegitimate ones, as shown in Table 7, we see that the
former compared with the latter yielded capsules, whether containing
many seeds or only a few, in the proportion of 77 to 45, or as 100 to 58.
But the inferiority of the illegitimate unions is here perhaps too great,
for on a subsequent occasion 100 long-styled and short-styled flowers
were illegitimately fertilized, and they together yielded 53 capsules:
therefore the rate of 77 to 53, or as 100 to 69, is a fairer one than that

of 100 to 58. / Returning to Table 7, if we consider only the good capsules, those from the two legitimate unions were to those from the two illegitimate in number as 71 to 31, or as 100 to 44. Again, if we take an equal number of capsules, whether good or bad, from the legitimately and illegitimately fertilized flowers, we find that the former contained seeds by weight compared with the latter as 50 to 24, or as 100 to 48; but if all the poor capsules are rejected, of which many were produced by the illegitimately fertilized flowers, the proportion is 54 to 35, or as 100 to 65. In this and all other cases, the relative fertility of the two kinds of union can, I think, be judged of more truly by the average number of seeds per capsule than by the proportion of flowers which yield capsules. The two methods might / have been combined by giving the average number of seeds produced by all the flowers which were fertilized, whether they yielded capsules or not; but I have thought that it would be more instructive always to show separately the proportion of flowers which produced capsules, and the average number of apparently good seeds which the capsules contained.

Flowers legitimately fertilized set seeds under conditions which cause the almost complete failure of illegitimately fertilized flowers. Thus in the spring of 1862 forty flowers were fertilized at the same time in both ways. The plants were accidentally exposed in the greenhouse to too hot a sun, and a large number of umbels perished. Some, however, remained in moderately good health, and on these there were twelve flowers which had been fertilized legitimately, and eleven which had been fertilized illegitimately. The twelve legitimate unions yielded seven fine capsules, containing on an average each 57·3 good seeds; whilst the eleven illegitimate unions yielded only two capsules, of which one contained 39 seeds, but so poor, that I do not suppose one would have germinated, and the other contained 17 fairly good seeds.

From the facts now given the superiority of a legitimate over an illegitimate union admits of not the least doubt; and we have here a case to which no parallel exists in the vegetable or, indeed, in the animal kingdom. The individual plants of the present species, and as we shall see of several other species of primula, are divided into two sets or bodies, which cannot be called distinct sexes, for both are hermaphrodites; yet they are to a certain extent sexually distinct, for they require reciprocal union for perfect fertility. As quadrupeds are divided into two nearly equal bodies of different sexes, / so here we have two bodies, approximately equal in number, differing in their sexual powers and related to each other like males and females. There are many hermaphrodite animals which cannot fertilize themselves, but must unite with another hermaphrodite. So it is with numerous plants; for the pollen is often mature and shed, or is mechanically protruded, before the flower's own stigma is ready; and such flowers absolutely require the presence of another hermaphrodite for sexual union. But with the cowslip and various other species of Primula there is this wide difference, that one individual, though it can fertilize itself imperfectly, must unite with another individual for full fertility, it cannot, however, unite with any other individual in the same manner as an hermaphrodite plant can unite with any other one of the same species; or as one snail or earth-worm can unite with any other hermaphrodite individual. On the contrary, an individual belonging to one form of the cowslip in order to be perfectly fertile must unite with one of the other form, just as a male quadruped must and can unite only with the female.

I have spoken of the legitimate unions as being fully fertile; and I am fully justified in doing so, for flowers artificially fertilized in this manner yielded rather more seeds than plants naturally fertilized in a state of nature. The excess may be attributed to the plants having been grown separately in good soil. With respect to the illegitimate union, we shall best appreciate their degree of lessened fertility by the following facts. Gärtner estimated the sterility of the unions between distinct species,[6] in a manner which allows of a strict comparison with the results of the / legitimate and illegitimate unions of Primula. With *P. veris*, for every 100 seeds yielded by the two legitimate unions, only 64 were yielded by an equal number of good capsules from the two illegitimate unions. With *P. Sinensis*, as we shall hereafter see, the proportion was nearly the same – namely, as 100 to 62. Now Gärtner has shown that, on the calculation of *Verbascum lychnitis* yielding with its own pollen 100 seeds, it yielded when fertilized by the pollen of *V. Phoeniceum* 90 seeds; by the pollen of *V. nigrum*, 63 seeds; by that of *V. blattaria*, 62 seeds. So again, *Dianthus barbatus* fertilized by the pollen of *D. superbus* yielded 81 seeds, and by the pollen of *D. Japonicus* 66 seeds, relatively to the 100 seeds produced by its own pollen. We thus see – and the fact is highly remarkable – that with Primula the illegitimate unions relatively to the legitimate are more sterile than crosses between distinct species of other genera relatively to their pure unions.

[6] *Versuche über die Bastarderzeugung*, 1849, p. 216.

Mr Scott has given[7] a still more striking illustration of the same fact: he crossed *Primula auricula* with pollen of four other species (*P. Palinuri, viscosa, hirsuta,* and *verticillata*), and these hybrid unions yielded a larger average number of seeds than did *P. auricula* when fertilized illegitimately with its own-form pollen.

The benefit which heterostyled dimorphic plants derive from the existence of the two forms is sufficiently obvious, namely, the inter-crossing of distinct plants being thus ensured.[8] Nothing can be better adapted for this end than the relative positions of the anthers and stigmas in the two forms, as shown in Fig. 2; but to / this whole subject I shall recur. No doubt pollen will occasionally be placed by insects or fall on the stigma of the same flower; and if cross-fertilization fails, such self-fertilization will be advantageous to the plant, as it will thus be saved from complete barrenness. But the advantage is not so great as might at first be thought, for the seedlings from illegitimate unions do not generally consist of both forms, but all belong to the parent form; they are, moreover, in some degree weakly in constitution, as will be shown in a future chapter. If, however, a flower's own pollen should first be placed by insects or fall on the stigma, it by no means follows that cross-fertilization will be thus prevented. It is well known that if pollen from a distinct species be placed on the stigma of a plant, and some hours afterwards its own pollen be placed on it, the latter will be prepotent and will quite obliterate any effect from the foreign pollen; and there can hardly be a doubt that with heterostyled dimorphic plants, pollen from the other form will obligate the effects of pollen from the same form, even when this has been placed on the stigma a considerable time before. To test this belief, I placed on several stigmas of a long-styled cowslip plenty of pollen from the same plant, and after twenty-four hours added some from a short-styled dark-red polyanthus, which is a variety of the cowslip. From the flowers thus treated 30 seedlings were raised, and all these, without exception, bore reddish flowers; so that the effect of pollen from the same form, though placed on the stigmas twenty-four hours previously, was quite destroyed by that of pollen from a plant belonging to the other form.

Finally, I may remark that that of the four kinds of unions, that of

---

[7] *Journ. Linn. Soc. Bot.,* vol. viii, 1864, p. 93

[8] I have shown in my work on the *Effects of Cross and Self-fertilisation* how greatly the offspring from intercrossed plants profit in height, vigour, and fertility.

the short-styled illegitimately fertilized with its own-form pollen seems to be the most sterile of / all, as judged by the average number of seeds, which the capsules contained. A smaller proportion, also, of these seeds than of the others germinated, and they germinated more slowly. The sterility of this union is the more remarkable, as it has already been shown that the short-styled plants yield a larger number of seeds than the long-styled, when both forms are fertilized, either naturally or artificially, in a legitimate manner.

In the future chapter, when I treat of the offspring from hetero-styled dimorphic and trimorphic plants illegitimately fertilized with their own-form pollen, I shall have occasion to show that with the present species and several others, equal-styled varieties sometimes appear.

PRIMULA ELATIOR, Jacq.
Bardfield oxlip of English authors

This plant, as well as the last or cowslip (*P. veris*, vel *officinalis*), and the primrose (*P. vulgaris*, vel *acaulis*) have been considered by some botanists as varieties of the same species. But they are all three undoubtedly distinct, as will be shown in the next chapter. The present species resembles to a certain extent in general appearance the common oxlip, which is a hybrid between the cowslip and primrose. *Primula elatior* is found in England only in two or three of the eastern counties; and I was supplied with living plants by Mr Doubleday, who, as I believe, first called attention to its existence in England. It is common in some parts of the Continent; and H. Müller[9] has seen several kinds of humble-bees and other bees, and Bombylius, visiting the flowers in North Germany. /

The results of my trials on the relative fertility of the two forms, when legitimately and illegitimately fertilized, are given in Table 8.

If we compare the fertility of the two legitimate unions taken together with that of the two illegitimate unions together, as judged by the proportional number of flowers which when fertilized in the two methods yielded capsules, the ratio is as 100 to 27; so that by this standard the present species is much more sterile than *P. veris*, when both species are illegitimately fertilized. If we judge of the relative

[9] *Die Befruchtung der Blumen*, p. 347.

23

TABLE 8 *Primula elatior*

Nature of union	Number of flowers fertilized	Number of good capsules produced	Maximum of seeds in any one capsule	Minimum of seeds in any one capsule	Average number of seeds per capsule
Long-styled form, by pollen of short-styled. Legitimate union	10	6	62	34	46·5
Long-styled form, by own-form pollen. Illegitimate union	20	4	49[10]	2	27·7
Short-styled form, by pollen of long-styled. Legitimate union	10	8	61	37	47·7
Short-styled form, by own-form pollen. Illegitimate union	17	3	19	9	12·1
The two legitimate unions together	20	14	62	37	47·1
The two illegitimate unions together	37	7	49[10]	2	35·5

fertility of the two kinds of unions by the average number of seeds per capsule, the ratio is as 100 to 75. But this latter / number is probably much too high, as many of the seeds produced by the illegitimately fertilized long-styled flowers were so small that they probably would not have germinated, and ought not to have been counted. Several long-styled and short-styled plants were protected from the access of insects, and must have been spontaneously self-fertilized. They yielded altogether only six capsules, containing any seeds; and their average number was only 7·8 per capsule. Some, moreover, of these seeds were so small that they could hardly have germinated.

Herr W. Breitenbach informs me that he examined, in two sites near the Lippe (a tributary of the Rhine), 894 flowers produced by 198 plants of this species; and he found 467 of these flowers to be long-styled, 411 short-styled, and 16 equal-styled. I have heard of no other instance with heterostyled plants of equal-styled flowers appearing in a state of nature, though far from rare with plants which have been long cultivated. It is still more remarkable that in eighteen cases the same plant produced both long-styled and short-styled, or long-styled and

[10] These seeds were so poor and small that they could hardly have germinated.

equal-styled flowers; and in two out of the eighteen cases, long-styled, short-styled, and equal-styled flowers. The long-styled flowers greatly preponderated on these eighteen plants – 61 consting of this form, 15 of equal-styled, and 9 of the short-styled form.

PRIMULA VULGARIS (var. *acaulis*, Linn.)
The primrose of English writers

Mr J. Scott examined 100 plants growing near Edinburgh, and found 44 to be long-styled, and 56 short-styled; and I took by chance 79 plants in Kent, of which 39 were long-styled and 40 short-styled; so / that the two lots together consisted of 83 long-styled and 96 short-styled plants. In the long-styled form the pistil is to that of the short-styled in length, from an average of five measurements, as 100 to 51. The stigma in the long-styled form is conspicuously more globose and much more papillose than in the short-styled, in which latter it is depressed on the summit; it is equally broad in the two forms. In both it stands nearly, but not exactly, on a level with the anthers of the opposite form; for it was found, from an average of 15 measurements, that the distance between the middle of the stigma and the middle of the anthers in the short-styled form is to that in the long-styled as 100

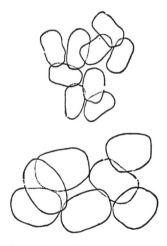

Fig. 3. Outlines of pollen grains of *Primula vulgaris*, distended with water, much magnified and drawn under the camera lucida. The upper and smaller grains from the long-styled form; the lower and larger grains from the short-styled.

to 93. The anthers do not differ in size in the two forms. The pollen grains from the short-styled / flowers before they were soaked in water were decidedly broader, in proportion to their length, than those from the long-styled; after being soaked they were relatively to those from the long-styled as 100 to 71 in diameter, and more transparent. A large number of flowers from the two forms were compared, and 12 of the finest flowers from each lot were measured, but there was no sensible difference between them in size. Nine long-styled and eight short-styled plants growing together in a state of nature were marked, and their capsules collected after they had been naturally fertilized; and the seeds from the short-styled weighed exactly twice as much as those from an equal number of long-styled plants. So that the primrose resembles the cowslip in the short-styled plants being the more productive of the two forms. The results of my trials on the fertility of the two forms, when legitimately and illegitimately fertilized, are given in Table 9.

TABLE 9   *Primula vulgaris*

Nature of union	Number of flowers fertilized	Number of good capsules produced	Maximum number of seeds in any one capsule	Minimum number of seeds in any one capsule	Average number of seeds per capsule
Long-styled form, by pollen from short-styled. Legitimate union	12	11	77	47	66·9
Long-styled form, by own-form pollen. Illegitimate union	21	14	66	30	52·2
Short-styled form, by pollen from long-styled. Legitimate union	8	7	75	48	65·0
Short-styled form, by own-form pollen. Illegitimate union	18	7	43	5	18·8[11]
The two legitimate unions together	20	18	77	47	66·0
The two illegitimate unions together	39	21	66	5	35·5[11]

[11] This average is perhaps rather too low.

We may infer from this table that the fertility of the two legitimate unions taken together is to that of the two illegitimate unions together, as judged by the proportional number of flowers which when fertilized in the two methods yielded capsules, as 100 to 60. If we judge by the average number of seeds per capsule produced by the two kinds of unions, the ratio is as 100 to 54; but this latter figure is perhaps rather too low. It is surprising how rarely insects can be seen during the day visiting the flowers, but I have occasionally observed small kinds of bees at work; I suppose, therefore, that they are commonly fertilized by nocturnal Lepidoptera. The long-styled plants when protected from insects yield a considerable number of capsules, and they thus differ remarkably from the same form of the cowslip, which is quite sterile under the same circumstances. Twenty-three spontaneously self-fertilized capsules from / this form contained, on an average, 19·2 seeds. The short-styled plants produced fewer spontaneously self-fertilized capsules, and fourteen of them contained only 6·2 seeds per capsule. The self-fertilization of both forms was probably aided by thrips, which abounded within the flowers; but these minute insects could not have placed nearly sufficient pollen on the stigmas, as the spontaneously self-fertilized capsules contained much fewer seeds, on an average, than those (as may be seen in Table 9) which were artificially fertilized with their own-form pollen. But this difference may perhaps be attributed in part to the flowers in the table having been fertilized with pollen from a distinct plant belonging / to the same form; whilst those which were spontaneously self-fertilized no doubt generally received their own pollen. In a future part of this volume some observations will be given on the fertility of a red-coloured variety of the primrose.

PRIMULA SINENSIS

In the long-styled form the pistil is about twice as long as that of the short-styled, and the stamens differ in a corresponding, but reversed, manner. The stigma is considerably more elongated and rougher than that of the short-styled, which is smooth and almost spherical, being somewhat depressed on the summit; but the stigma varies much in all its characters, the result, probably, of cultivation. The pollen grains of the short-styled form, according to Hildebrand,[12] are 7 divisions of the micrometer in length and 5 in breadth; whereas those of the long-styled are only 4 in length and 3 in breadth. The grains, therefore, of the short-styled are to those of the long-styled in length as 100 to 57. Hildebrand also remarked, as I had done in the case of *P. veris*, that the smaller grains from the long-styled are much more transparent than the larger ones from the short-styled form. We shall hereafter see that this cultivated plant varies much in its dimorphic condition and is often equal-styled. Some individuals may be said to be sub-heterostyled; thus in two white-flowered plants the pistil projected above the stamens, but in one of them / it was longer and had a more elongated and rougher stigma, than in the other; and the pollen grains from the latter were to those from the plant with a more elongated pistil only as 100 to 88 in diameter, instead of as 100 to 57. The corolla of the long-styled and short-styled form differs in shape, in the same manner as in *P. veris*. The long-styled plants tend to flower before the short-styled. When both forms were legitimately fertilized, the capsules from the short-styled plants contained, on an average, more seeds than those from the long-styled, in the ratio of 12·2 to 9·3 by weight, that is, as 100 to 78. In Table 10 we have the results of two sets of experiments tried at different periods. /

The fertility, therefore, of the two legitimate unions together to that of the two illegitimate unions, as judged by the proportional number of flowers which yielded capsules, is as 100 to 84. Judging by the average weight of seeds per capsule produced by the two kinds of unions, the ratio is as 100 to 63. On another occasion a large number of flowers of both forms were fertilized in the same manner, but no account of their number was kept. The seeds, however, were carefully counted, and the averages are shown in the right-hand column. The ratio for the number of seeds produced by the two legitimate compared with the two illegitimate unions is here 100 to 53, which is probably more accurate than the foregoing one of 100 to 63.

Hildebrand in the paper above referred to gives the results of his experiments on the present species; and these are shown in a condensed form in Table 11. Besides using for the illegitimate unions pollen from a distinct plant of the same form, as was always done by

[12] After the appearance of my paper this author published some excellent observations on the present species (*Bot. Zeitung*, 1 January, 1864), and he shows that I erred greatly about the size of the pollen grains in the two forms. I suppose that by mistake I measured twice over pollen grains from the same form.

TABLE 10.   *Primula Sinensis*

Nature of union	Number of flowers fertilized	Number of good capsules produced	Average weight of seeds per capsule	Average number of seeds per capsule, as ascertained on a subsequent occasion
Long-styled form, by pollen of short-styled. Legitimate union	24	16	0·58	50
Long-styled form, by own-form pollen. Illegitimate union	20	13	0·45	35
Short-styled form, by pollen of long-styled. Legitimate union	8	8	0·76	64
Short-styled form, by own-form pollen. Illegitimate union	7	4	0·23	25
The two legitimate unions together	32	24	0·64	57
The two illegitimate unions together	27	17	0·40	30

TABLE 11.   *Primula Sinensis (from Hildebrand)*

Nature of union	Number of flowers fertilized	Number of good capsules produced	Average number of seeds per capsule
Long-styled form, by pollen of short-styled. Legitimate union	14	14	41
Long-styled form, by own-form pollen, from a distinct plant. Illegitimate union	26	26	18
Long-styled form, by pollen from same flower. Illegitimate union	27	21	17
Short-styled form by pollen of long-styled. Legitimate union	14	14	44
Short-styled form, by own-form pollen, from a distinct plant. Illegitimate union	16	16	20
Short-styled, by pollen from the same flower. Illegitimate union	21	11	8
The two legitimate unions together	28	28	43
The two illegitimate unions together (own-form pollen)	42	42	18
The two illegitimate unions together (pollen from the same flower)	48	32	13

me, he tried, in addition, the effects of the plant's own pollen. He counted the seeds.

It is remarkable that here all the flowers which were fertilized legitimately, as well as those fertilized illegitimately with pollen from a distinct plant belonging to the same form, yielded capsules; and from this fact it might be inferred that the two forms were reciprocally much more fertile in his case than in mine. But his illegitimately fertilized capsules from both forms contained fewer seeds relatively to the legitimately fertilized capsules than in my experiments; for the ratio in his case is as 42 to 100, instead of, as in mine, as 53 to 100. Fertility is a very variable element with most plants, being determined by the conditions to which they are subjected, of which fact I have observed striking instances with the / present species; and this may account for the difference between my results and those of Hildebrand. His plants were kept in a room, and perhaps were grown in too small pots or under some other unfavourable conditions, for his capsules in almost every case contained a smaller number of seeds than mine, as may be seen by comparing the right-hand columns in Tables 10 and 11.

The most interesting point in Hildebrand's experiments is the difference in the effects of illegitimate fertilization with a flower's own pollen, and with that / from a distinct plant of the same form. In the latter case all the flowers produced capsules, whilst only 67 out of 100 of those fertilized with their own pollen produced capsules. The self-fertilized capsules also contained seeds, as compared with capsules from flowers fertilized with pollen from a distinct plant of the same form, in the ratio of 72 to 100.

In order to ascertain how far the present species was spontaneously self-fertile, five long-styled plants were protected by me from insects; and they bore up to a given period 147 flowers which set 62 capsules; but many of these soon fell off, showing that they had not been properly fertilized. At the same time five short-styled plants were similarly treated, and they bore 116 flowers which ultimately produced only seven capsules. On another occasion 13 protected long-styled plants yielded by weight 25·9 grains of spontaneously self-fertilized seeds. At the same time seven protected short-styled plants yielded only half-a-grain weight of seeds. Therefore the long-styled plants yielded nearly 24 times as many spontaneously self-fertilized seeds as did the same number of short-styled plants. The chief cause of this great difference appears to be, that when the corolla of a long-styled plant falls off, the anthers, from being situated near the bottom of the tube, are necessarily dragged over the stigma and leave pollen on it, as I saw when I hastened the fall of nearly withered flowers; whereas, in the short-styled flowers, the stamens are seated at the mouth of the corolla, and in falling off do not brush over the lowly-seated stigmas. Hildebrand likewise protected some long-styled and short-styled plants, but neither ever yielded a single capsule. He thinks that the difference in our results may be accounted for by his plants having been kept in a room and never having been shaken; / but this explanation seems to me doubtful; his plants were in a less fertile condition than mine, as shown by the difference in the number of seeds produced, and it is highly probable that their lessened fertility would have interfered with especial force with their capacity for producing self-fertilized seeds.

PRIMULA AURICULA

This species is heterostyled, like the preceding ones; but among the varieties distributed by florists the long-styled form is rare, as it is not valued. There is a much greater relative inequality in the length of the pistil and stamens in the two forms of the auricula than in the cowslip; the pistil in the long-styled being nearly four times as long as that in the short-styled, in which it is barely longer than the ovarium. The stigma is nearly of the same shape in both forms, but is rougher in the long-styled, though the difference is not so great as between the two forms of the cowslip. In the long-styled plants the stamens are very short, rising but little above the ovarium. The pollen-grains of these short stamens, when distended with water, were barely $\frac{5}{6000}$ of an inch in diameter, whereas those from the long stamens of the short-styled plants were barely $\frac{7}{6000}$, showing a relative difference of about 71 to 100. The smaller grains of the long-styled plant are also much more transparent, and before distension with water more triangular in outline than those of the other form. Mr Scott[14] compared ten plants of both forms growing under similar conditions, and found that, although the long-styled plants produced more umbels and more capsules than the short-styled, yet they yielded fewer seeds, in the ratio of 66 to 100. Three short-styled plants were protected by me from the / access of insects, and they did not produce a single seed. Mr Scott protected six plants of both forms, and found them excessively sterile. The pistil of the long-styled

[13] According to Kerner, our garden auriculas are descended from *P. pubescens*, Jacq., which is a hybrid between the true *P. auricula* and *hirsuta*. This hybrid has now been propagated for about 300 years, and produces, when legitimately fertilized, a large number of seeds; the long-styled forms yielding an average number of 73, and the short-styled 98 seeds per capsule: see his 'Geschichte der Aurikel', *Zeitschr. des Deutschen und Oest. Alpen-Vereins*, vol. vi, p. 52. Also 'Die Primulaceen-Bastarten', *Oest. Bot. Zeitschrift*, 1835, nos 3, 4, and 5.
[14] *Journ. Linn. Soc. Bot.*, vol. viii, 1864, p. 86.

31

form stands so high above the anthers, that it is scarcely possible that pollen should reach the stigma without some aid; and one of Mr Scott's long-styled plants which yielded a few seeds (only 18 in number) was infested by aphides, and he does not doubt that these had imperfectly fertilized it.

I tried a few experiments by reciprocally fertilizing the two forms in the same manner as before, but my plants were unhealthy, so I will give, in a condensed form, the results of Mr Scott's experiments. For fuller particulars with respect to this and the five following species, the paper lately referred to may be consulted. In each case the fertility of the two legitimate unions, taken together, is compared with that of the two illegitimate unions together, by the same two standards as before, namely, by the proportional number of flowers which produced good capsules, and by the average number of seeds per capsule. The fertility of the legitimate unions is always taken at 100.

By the first standard, the fertility of the two legitimate unions of the auricula is to that of the two illegitimate unions as 100 to 80; and by the second standard as 100 to 15.

PRIMULA SIKKIMENSIS

According to Mr Scott the pistil of the long-styled form is fully four times as long as that of the short-styled, but their stigmas are nearly alike in shape and roughness. The stamens do not differ so much in relative length as the pistils. The pollen grains differ in a marked manner in the two forms; 'those of the long-styled plants are sharply triquetrous, smaller, and more transparent than those of the short-styled, which are of a bluntly triangular form'. The fertility of the two legitimate unions to that of the two illegitimate unions is by the first standard as 100 to 95, and by the second standard as 100 to 31.

PRIMULA CORTUSOIDES

The pistil of the long-styled form is about thrice as long as that of the short-styled, the stigma being double as long and covered with much longer papillae. The pollen grains of the short-styled / form are, as usual, 'larger, less transparent, and more bluntly triangular than those from the long-styled plants'. The fertility of the two legitimate unions to that of the two illegitimate unions is by the first standard as 100 to 74, and by the second standard as 100 to 66.

PRIMULA INVOLUCRATA

The pistil of the long-styled form is about thrice as long as that of the short-styled; the stigma of the former is globular and closely beset with papillae, whilst that of the short-styled is smooth and depressed on the apex. The pollen grains of the two forms differ in size and transparency as before, but not in shape. The fertility of the two legitimate to that of the two illegitimate unions is by the first standard as 100 to 72; and by the second standard as 100 to 47.

PRIMULA FARINOSA

According to Mr Scott, the pistil of the long-styled form is only about twice as long as that of the short-styled. The stigmas of the two forms differ but little in shape. The pollen grains differ in the usual manner in size, but not in form. The fertility of the two legitimate to that of the two illegitimate unions is by the first standard as 100 to 71, and by the second standard as 100 to 44.

Summary on the foregoing heterostyled species of Primula

The fertility of the long- and short-styled plants of the above species of Primula, when the two forms are fertilized legitimately, and illegitimately with pollen of the same form taken from a distinct plant, has now been given. The results are seen in the following table; the fertility being judged by two standards, namely, by that of the proportional number of flowers which yielded capsules, and by that of the average number of seeds per capsule. But for full accuracy many more observations, under varied conditions, would be requisite. /

TABLE 12

Summary on the fertility of the two legitimate unions, compared with that of the two illegitimate unions, in the genus Primula. The former taken at 100

| | ILLEGITIMATE UNIONS | |
Name of Species	Judged of by the proportional number of flowers which produced capsules	Judged of by the average number (or weight in some cases) of seeds per capsule
Primula veris	69	65
P. elatior	27	75 (probably too high)
P. vulgaris	60	54 (perhaps too low)
P. Sinensis	84	63
P. Sinensis (second trial)	?	53
P. Sinensis (after Hildebrand)	10	42
P. auricula (Scott)	80	15
P. Sikkimensis (Scott)	95	31
P. cortusoides (Scott)	74	66
P. involucrata (Scott)	72	48
P. farinosa (Scott)	71	44
Average of the nine species	88·4	61·8

With plants of all kinds some flowers generally fail to produce capsules, from various accidental causes; but this source of error has been eliminated, as far as possible, in all the previous cases, by the manner in which the calculations have been made. Supposing, for instance, that 20 flowers were fertilized legitimately and yielded 18 capsules, and that 30 flowers were fertilized illegitimately and yielded 15 capsules, we may assume that on an average an equal proportion of the flowers in both lots would fail to produce capsules from various accidental causes; and the ratio of $^{18}/_{20}$ to $^{15}/_{30}$, or as 100 to 56 (in whole / numbers), would show the proportional number of capsules due to the two methods of fertilization; and the number 56 would appear in the left-hand column of Table 12, and in any other tables. With respect to the average number of seeds per capsule hardly anything need be said: supposing that the legitimately fertilized capsules contained, on an average, 50 seeds, and the illegitimately fertilized capsules 25 seeds; then as 50 is to 25 so is 100 to 50; and the latter number would appear in the right-hand column.

It is impossible to look at the above table and doubt that the legitimate unions between the two forms of the above nine species of Primula are much more fertile than the illegitimate unions; although in the latter case pollen was always taken from a distinct plant of the same form. There is, however, no close correspondence in the two rows of figures, which give, according to the two standards, the difference of fertility between the legitimate and illegitimate unions. Thus all the flowers of *P. Sinensis* which were illegitimately fertilized by Hildebrand produced capsules; but these contained only 42 per cent of the number of seeds yielded by the legitimately fertilized capsules. So again, 95 per cent of the illegitimately fertilized flowers of *P. Sikkimensis* produced capsules; but these contained only 31 per cent of the number of seeds in the legitimate capsules. On the other hand, with *P. elatior* only 27 per cent of the illegitimately fertilized flowers yielded capsules; but these contained nearly 75 per cent of the legitimate number of seeds. It appears that the setting of the flowers, that is, the production of capsules whether good or bad, is not so much influenced by legitimate and illegitimate fertilization as is the number of seeds which the capsules / contain. For, as may be seen at the bottom of Table 12, 88·4 per cent of the illegitimately fertilized flowers yielded capsules; but these contained only 61·8 per cent of seeds, in comparison, in each case, with the legitimately fertilized flowers and capsules of the same species.

There is another point which deserves notice, namely, the relative degree of infertility in the several species of the long-styled and short-styled flowers, when both are illegitimately fertilized. The data may be found in the earlier tables, and in those given by Mr Scott in the paper already referred to. If we call the number of seeds per capsule produced by the illegitimately fertilized long-styled flowers 100, the seeds from the illegitimately fertilized short-styled flowers will be represented by the following numbers:

Primula veris	71	Primula auricula	119
P. elatior	44 (probably too low)	P. Sikkimensis	57
P. vulgaris	36 (perhaps too low)	P. cortusoides	93
P. Sinensis	71	P. involucrata	74
		P. farinosa	63

We thus see that, with the exception of *P. auricula*, the long-styled flowers of all nine species are more fertile than the short-styled flowers, when both forms are illegitimately fertilized. Whether *P. auricula* really differs from the other species in this respect I can form no opinion, as the result may have been accidental. The degree of self-fertility of a plant depends on two elements, namely, on the stigma receiving its own pollen and on its more or less efficient action when placed there. Now as the anthers of the short-styled flowers of several species of Primula stand directly above the stigma, their pollen is more likely to fall on it, or to be carried down to it by insects, than in the case of / the long-styled form. It appears probable, therefore at first sight, that the lessened capacity of the short-styled flowers to be fertilized with their own pollen, is a special adaptation for counteracting their greater liability to receive their own pollen, and thus for checking self-fertilization. But from facts with respect to other species hereafter to be given, this view can hardly be admitted. In accordance with the above liability, when some of the species of Primula were allowed to fertilize themselves spontaneously under a net, all insects being excluded, except such minute ones as thrips, the short-styled flowers, notwithstanding their greater innate self-fertility, yielded more seed than did the long-styled. None of the species, however, when insects were excluded, made a near approach to full fertility. But the long-styled form of *P. Sinensis* gave, under these circumstances, a considerable number of seeds, as the corolla in falling off drags the anthers, which are seated low down in the tube, over the stigma, and thus leaves plenty of pollen on it.

35

Homostyled species of Primula. It has now been shown that nine of the species in this genus exist under two forms, which differ not only in structure but in function. Besides these Mr Scott enumerates 27 other species[15] which are heterostyled; and to these probably others will be hereafter added. Nevertheless, some species are homostyled; that is, they exist only under a single form; but much caution is necessary on this head, as several species when cultivated are apt to become equal-styled. Mr Scott believes that *P. Scotica, verticillata,* a variety of *Sibirica, elata, mollis,* and / *longiflora,*[16] are truly homostyled; and to these may be added, according to Axell, *P. stricta.* Mr Scott experimented on *P. Scotica, mollis,* and *verticillata,* and found that their flowers yielded an abundance of seeds when fertilized with their own pollen. This shows that they are not heterostyled in function. *P. Scotica* is, however, only moderately fertile when insects are excluded, but this depends merely on the coherent pollen not readily falling on the stigma without their aid. Mr Scott also found that the capsules of *P. verticillata* contained rather more seed when the flowers were fertilized with pollen from a distinct plant than when with their own pollen; and from this fact he infers that they are sub-heterostyled in function, though not in structure. But there is no evidence that two sets of individuals exist, which differ slightly in function and are adapted for reciprocal fertilization; and this is the essence of heterostylism. The mere fact of a plant being more fertile with pollen from a distinct individual than with its own pollen, is common to very many species, as I have shown in my work *On the Effects of Cross and Self-fertilization.*

HOTTONIA PALUSTRIS

This aquatic member of the Primulaceae is conspicuously heterosty-led, as the pistil of the long-styled form projects far out of the flower, the stamens being enclosed within the tube; whilst the stamens of the short-styled flower project far forwards, the pistil being enclosed. This difference between the two forms has attracted the attention of

[15] H. Müller has given in *Nature,* 10 December, 1874, p. 110, a drawing of one of these species, viz. the alpine *P. villosa,* and shows that it is fertilized exclusively by Lepidoptera.

[16] Koch was aware that this species was homostyled: see 'Treviranus über Dichogamie nach Sprengel und Darwin', *Bot. Zeitung,* 2 January, 1863, p. 4.

various botanists, and that / of Sprengel,[17] in 1793, who, with his usual sagacity, adds that he does not believe the existence of the two forms to be accidental, though he cannot explain their purpose. The pistil of the long-styled form is more than twice as long as that of the short-styled, with the stigma rather smaller, though rougher. H. Müller[18] gives figures of the stigmatic papillae of the two forms, and those of the long-styled are seen to be more than double the length, and much thicker than the papillae of the short-styled form. The anthers in the one form do not stand exactly on a level with the stigma in the other form; for the distance between the organs is greater in the short-styled than in the long-styled flowers in the proportion of 100 to 71. In dried specimens soaked in water the anthers of the short-styled form are larger than those of the long-styled, in the ratio of 100 to 83. The pollen grains, also, from the short-styled flowers are conspicuously larger than those from the long-styled; the ratio between the diameters of the moistened grains being as 100 to 64, according to my measurements, but according to the measurements of H. Müller as 100 to 61; and his are probably the more accurate of the two. The contents of the larger pollen grains appear more coarsely granular and of a browner tint, than those in the smaller grains. The two forms of Hottonia thus agree closely in most respects with those of the heterostyled species of Primula. The flowers of Hottonia are cross-fertilized, according to Müller, chiefly by Diptera.

Mr Scott[19] made a few trials on a short-styled plant, and found that the legitimate unions were in all ways more fertile than the illegitimate; but since the publication / of his paper H. Müller has made much fuller experiments, and I give his results in Table 13, drawn up in accordance with my usual plan.

The most remarkable point in this table is the small average number of seeds from the short-styled flowers when illegitimately fertilized, and the unusually large average number of seeds yielded by the illegitimately fertilized long-styled flowers, relatively in both cases to the product of the legitimately fertilized flowers.[20] The two legitimate

---

[17] Das entdeckte Geheimniss der Natur, p. 103.

[18] Die Befruchtung, etc., p. 350.

[19] Journ. Linn. Soc. Bot., vol. viii, 1864, p. 79.

[20] H. Müller says (Die Befruchtung, etc., p. 352) that the long-styled flowers, when illegitimately fertilized, yield as many seeds as when legitimately fertilized; but by adding up the number of seeds from all the capsules produced by the two methods of fertilization, as given by him, I arrive at the results shown in Table 13. The average number in the long-styled capsules, when legitimately fertilized, is 91·4, and when illegitimately fertilized, 77·5; or as 100 to 85. H. Müller agrees with me that this is the proper manner of viewing the case.

TABLE 13
Hottonia palustris (from H. Müller)

Nature of union	Number of capsules examined	Average number of seeds per capsule
Long-styled form, by pollen of short-styled. Legitimate union	34	91·4
Long-styled form, by own-form pollen, from a distinct plant. Illegitimate union	18	77·5
Short-styled form, by pollen of long-styled. Legitimate union	30	66·2
Short-styled form, by own-form pollen, from a distinct plant. Illegitimate union	19	18·7
The two legitimate unions together	64	78·8
The two illegitimate unions together	37	48·1

unions compared with / the two illegitimate together yield seeds in the ratio of 100 to 61.

H. Müller also tried the effects of illegitimately fertilizing the long-styled and short-styled flowers with their own pollen, instead of with that from another plant of the same form; and the results are very striking. For the capsules from the long-styled flowers thus treated contained, on an average, only 15·7 seeds instead of 77·5; and those from the short-styled 6·5, instead of 18·7 seeds per capsule. The number 6·5 agrees closely with Mr Scott's result from the same form similarly fertilized.

From some observations by Dr Torrey, *Hottonia inflata*, an inhabitant of the United States, does not appear to be heterostyled, but is remarkable from producing cleistogamic flowers, as will be seen in the last chapter of this volume.

Besides the general Primula and Hottonia, *Androsace* (vel Gregoria, vel Aretia) *vitalliana* is heterostyled, Mr Scott[21] fertilized with their own pollen 21 flowers on three short-styled plants in the Edinburgh Botanic Gardens, and not one yielded a single seed; but eight of them which were fertilized with pollen from one of the other plants of the same form, set two empty capsules. He was able to examine only dried specimens of the long-styled forms. But the evidence seems sufficient to leave hardly a doubt that Androsace is heterostyled. Fritz Müller sent me from South Brazil dried flowers of a Statice which he believed

[21] See also Treviranus in *Bot. Zeitung*, 1863, p. 6, on this plant being dimorphic.

to be heterostyled. In the one form the pistil was considerably longer and the stamens slightly shorter than the corresponding organs in the other form. But as in the shorter-styled form the stigmas reached up to the anthers / of the same flower, and as I could not detect in the dried specimens of the two forms any difference in their stigmas, or in the size of their pollen grains, I dare not rank this plant as heterostyled. From statements made by Vaucher I was led to think that *Soldanella alpina* was heterostyled, but it is impossible that Kerner, who has closely studied this plant, could have overlooked the fact. So again from other statements it appeared probable that Pyrola might be heterostyled, but H. Müller examined for me two species in North Germany, and found this not to be the case. /

CHAPTER II

HYBRID PRIMULAS

The oxlip a hybrid naturally produced between *Primula veris* and *vulgaris* – The differences in structure and function between the two parent species – Effects of crossing long-styled and short-styled oxlips with one another and with the two forms of both parent species – Character of the offspring from oxlips artificially self-fertilized and cross-fertilized in a state of nature – *Primula elatior* shown to be a distinct species – Hybrids between other heterostyled species of primula – Supplementary note on spontaneously produced hybrids in the genus Verbascum.

The various species of primula have produced in a state of nature throughout Europe an extraordinary number of hybrid forms. For instance, Professor Kerner has found no less than twenty-five such forms in the Alps.[1] The frequent occurrence of hybrids in this genus no doubt has been favoured by most of the species being heterostyled, and consequently requiring cross-fertilization by insects; yet in some other genera, species which are not heterostyled and which in some respects appears not well adapted for hybrid-fertilization, have likewise been largely hybridized. In certain districts of England, the common oxlip – a hybrid between the cowslip (*P. veris*, vel *officinalis*) and the primrose (*P. vulgaris*, vel *acaulis*) – is frequently found, and it occurs occasionally almost everywhere. / Owing to the frequency of this intermediate hybrid form, and to the existence of the Bardfield oxlip (*P. elatior*), which resembles to a certain extent the common oxlip, the claim of the three forms to rank as distinct species has been discussed oftener and at greater length than that of almost any other plant. Linnaeus considered *P. veris, vulgaris* and *elatior* to be varieties of the same species, as do some distinguished botanists at the present day; whilst others who have carefully studied these plants do not doubt that they are distinct species. The following observations prove, I think,

[1] 'Die Primulaceen-Bastarten', *Oesterr. Bot. Zeitschrift*, Jahr, 1875, nos 3, 4, and 5. See also Godron on hybrid primulas in *Bull. Soc. Bot. de France*, vol. x, 1853, p. 178. Also in *Revue des Sciences Nat.*, 1875, p. 331.

that the latter view is correct; and they further show that the common oxlip is a hybrid between *P. veris* and *vulgaris*.

The cowslip differs so conspicuously in general appearance from the primrose, that nothing need here be said with respect to their external characters.[2] But some less obvious diferences deserve notice. As both species are heterostyled, their complete fertilization depends on insects. The cowslip is habitually visited during the day by the larger humble-bees (viz. *Bombus muscorum* and *hortorum*), and at night by moths, as I have seen in the case of *Cucullia*. The primrose is never visited (and I speak after many years' observation) by the larger humble-bees, and only rarely by the smaller kinds; hence its fertilization must depend almost exclusively on moths. There is nothing in the structure of the flowers of the two plants which can determine the visits of such widely different insects. But they emit a different odour, and perhaps their nectar may have a different taste. Both the long-styled and short-styled forms of / the primrose, when legitimately and naturally fertilized, yield on an average many more seeds per capsule than the cowslip, namely, in the proportion of 100 to 55. When illegitimately fertilized they are likewise more fertile than the two forms of the cowslip, as shown by the larger proportion of their flowers which set capsules, and by the larger average number of seeds which the capsules contain. The difference also between the number of seeds produced by the long-styled and short-styled flowers of the primrose, when both are illegitimately fertilized, is greater than that between the number produced under similar circumstances by the two forms of the cowslip. The long-styled flowers of the primrose when protected from the access of all insects, except such minute ones as thrips, yield a considerable number of capsules containing on an average 19·2 seeds per capsule; whereas 18 plants of the long-styled cowslip similarly treated did not yield a single seed.

The primrose, as every one knows, flowers a little earlier in the spring than the cowslip, and inhabits slightly different stations and districts. The primrose generally grows on banks or in woods, whilst the cowslip is found in more open places. The geographical range of the two forms is different. Dr Bromfield remarks[3] that 'the primrose is absent from all the interior region of northern Europe, where the

---

[2] The Rev. W. A. Leighton has pointed out certain differences in the form of the capsules and seed, in *Ann. and Mag. of Nat. Hist.*, 2nd series, vol. ii, 1848, p. 164.

[3] *Phytologist*, vol. iii, p. 694.

41

cowslip is indigenous'. In Norway, however, both plants range to the same degree of north latitude.[4]

The cowslip and primrose, when intercrossed, behave / like distinct species, for they are far from being mutually fertile. Gärtner[5] crossed 27 flowers of *P. vulgaris* with pollen of *P. veris*, and obtained 19 capsules; but these did not contain any good seed. He also crossed 21 flowers of *P. veris* with pollen of *P. vulgaris*; and now he got only five capsules, containing seed in a still less perfect condition. Gärtner knew nothing about heterostylism; and his complete failure may perhaps be accounted for by his having crossed together the same forms of the cowslip and primrose; for such crosses would have been of an illegitimate as well as of a hybrid nature, and this would have increased their sterility. My trials were rather more fortunate. Twenty-one flowers, consisting of both forms of the cowslip and primrose, were intercrossed legitimately, and yielded seven capsules (i.e. 33 per cent), containing on an average 42 seeds; some of these seeds, however, were so poor that they probably would not have germinated. Twenty-one flowers on the same cowslip and primrose plants were also intercrossed illegitimately, and they likewise yielded seven capsules (or 33 per cent), but these contained on an average only 13 good and bad seeds. I should, however, state that some of the above flowers of the primrose were fertilized with pollen from the polyanthus, which is certainly a variety of the cowslip, as may be inferred from the perfect fertility *inter se* of the crossed offspring from these two plants.[6] To show how sterile these hybrid unions / were, I may remind the reader that 90 per cent of the flowers of the primrose fertilized legitimately with primrose pollen yielded capsules, containing on an average 66 seeds; and that 54 per cent of the flowers fertilized illegitimately yielded capsules

---

[4] H. Lecoq, *Géograph. Bot. de l'Europe*, vol. viii, 1858, pp. 141, 144. See also *Ann. and Mag. of Nat. Hist.*, ix, 1842, pp. 156, 515. Also Boreau, *Flore du centre de la France*, 1840, vol. ii, p. 376. With respect to the rarity of *P. veris* in western Scotland, see H. C. Watson, *Cybele Britannica*, ii, p. 293.

[5] *Bastarderzeugung*, 1849, p. 721.

[6] Mr Scott has discussed the nature of the polyanthus (*Proc. Linn. Soc.*, viii, Bot., 1864, p. 103), and arrives at a different conclusion; but I do not think that his experiments were sufficiently numerous. The degree of infertility of a cross is liable to much fluctuation. Pollen from the cowslip at first appears rather more efficient on the primrose than that of the polyanthus; for 12 flowers of both forms of the primrose, fertilized legitimately and illegitimately with pollen of the cowslip gave five capsules containing on an average 32·4 seeds; whilst 18 flowers similarly fertilized by polyanthus pollen yielded only five capsules, containing only 22·6 seeds. On the other hand, the seeds produced by the polyanthus pollen were much the finest of the whole lot, and were the only ones which germinated.

containing on an average 3·55 seeds per capsule. The primrose, especially the short-styled form, when fertilized by the cowslip, is less sterile, as Gärtner likewise observed, than is the cowslip when fertilized by the primrose. The above experiments also show that a cross between the same forms of the primrose and cowslip is much more sterile than that between different forms of these two species.

The seeds from the several foregoing crosses were sown, but none germinated except those from the short-styled primrose fertilized with pollen of the polyanthus; and these seeds were the finest of the whole lot. I thus raised six plants, and compared them with a group of wild oxlips which I had transplanted into my garden. One of these wild oxlips produced slightly larger flowers than the others, and this one was identical in every character (in foliage, flower-peduncle, and flowers) with my six plants, excepting that the flowers of the latter were tinged of a dingy red colour, from being descended from the polyanthus.

We thus see that the cowslip and primrose cannot be crossed either way except with considerable difficulty, that they differ conspicuously in external appearance, that they differ in various physiological / characters, that they inhabit slightly different stations and range differently. Hence those botanists who rank these plants as varieties ought to be able to prove that they are not as well fixed in character as are most species; and the evidence in favour of such instability of character appears at first sight very strong. It rests, first, on statements made by several competent observers that they have raised cowslips, primroses, and oxlips from seeds of the same plant; and, secondly, on the frequent occurrence in the state of nature of plants presenting every intermediate gradation between the cowslip and primrose.

The first statement, however, is of little value; for, heterostylism not being formerly understood, the seed-bearing plants were in no instance[7] protected from the visits of insects; and there would be almost as much risk of an isolated cowslip, or of several cowslips if consisting of the same form, being crossed by a neighbouring primrose and producing oxlips, as of one sex of a dioecious plant, under similar

---

[7] One author states in the *Phytologist* (vol. iii, p. 703) that he covered with bell-glasses some cowslips, primroses, etc., on which he experimented. He specifies all the details of his experiment, but does not say that he artificially fertilized his plants; yet he obtained an abundance of seed, which is simply impossible. Hence there must have been some strange error in these experiments, which may be passed over as valueless.

circumstances, being crossed by the opposite sex of an allied and neighbouring species. Mr H. C. Watson, a critical and most careful observer, made many experiments by sowing the seeds of cowslips and of various kinds of oxlips, and arrived at the following conclusion,[8] namely, 'that seeds of the cowslip can produce cowslips and oxlips, and that seeds of an oxlip can produce cowslips, oxlips, and primroses'. This conclusion harmonises perfectly with the view that in / all cases, when such results have been obtained, the unprotected cowslips have been crossed by primroses, and the unprotected oxlips by either cowslips or primroses; for in this latter case we might expect, by the aid of reversion, which notoriously comes into powerful action with hybrids, that the two parent forms in appearance pure, as well as many intermediate gradations, would be occasionally produced. Nevertheless the two following statements offer considerable difficulty. The Rev. Professor Henslow[9] raised from seeds of a cowslip growing in his garden, various kinds of oxlips and one perfect primrose; but a statement in the same paper perhaps throws light on this anomalous result. Professor Henslow had previously transplanted into his garden a cowslip, which completely changed its appearance during the following year, and now resembled an oxlip. Next year again it changed its character, and produced, in addition to the ordinary umbels, a few single-flowered scapes, bearing flowers somewhat smaller and more deeply coloured than those of the common primrose. From what I have myself observed with oxlips, I cannot doubt that this plant was an oxlip in a highly variable condition, almost like that of the famous *Cytisus adami*. This presumed oxlip was propagated by offsets, which were planted in different parts oif the garden; and if Professor Henslow took by mistake seeds from one of these plants, especially if it had been crossed by a primrose, the result would be quite intelligible. Another case is still more difficult to understand: Dr Herbert[10] raised, from the seeds of a highly cultivated red cowslip, cowslips, oxlips of various kinds, and a primrose. This case, if accurately / recorded, which I must doubt, is explicable only on the improbable assumption that the red cowslip was not of pure parentage. With species and varieties of many kinds, when intercrossed, one is sometimes strongly prepotent

---

[8] *Phytologist*, ii, pp. 217, 852; iii, p. 43.
[9] Loudon's *Mag. of Nat. Hist.*, iii, 1830, p. 409.
[10] *Transact. Hort. Soc.*, iv, p. 19

over the other; and instances are known[11] of a variety, crossed by another, producing offspring which in certain characters, as in colour, hairiness, etc., have proved identical with the pollen-bearing parent, and quite dissimilar to the mother-plant; but I do not know of any instance of the offspring of a cross perfectly resembling, in a considerable number of important characters, the father alone. It is, therefore, very improbable that a pure cowslip crossed by a primrose should ever produce a primrose in appearance pure. Although the facts given by Dr Herbert and Professor Henslow are difficult to explain, yet until it can be shown that a cowslip or a primrose, carefully protected from insects, will give birth to at least oxlips, the cases hitherto recorded have little weight in leading us to admit that the cowslip and primrose are varieties of one and the same species.

Negative evidence is of little value; but the following facts may be worth giving: Some cowslips which had been transplanted from the fields into a shrubbery were again transplanted into highly manured land. In the following year they were protected from insects, artificially fertilized, and the seed thus procured was sown in a hotbed. The young plants were afterwards planted out, some in very rich soil, some in stiff poor clay, some in old peat, and some in pots in the greenhouse; so that these plants, 765 in number, as well as their parents, were subjected to diversified and unnatural / treatment; but not one of them presented the least variations except in size – those in the peat attaining almost gigantic dimensions, and those in the day being much dwarfed.

I do not, of course, doubt that cowslips exposed during *several* successive generations to changed conditions would vary, and that this might occasionally occur in a state of nature. Moreover, from the law of analogical variation, the varieties of any one species of Primula would probably in some cases resemble other species of the genus. For instance, I raised a red primrose from seed from a protected plant, and the flowers, though still resembling those of the primrose, were born during one season in umbels on a long footstalk like that of a cowslip.

With regard to the second class of facts in support of the cowslip and primrose being ranked as mere varieties, namely, the well-ascertained existence in a state of nature of numerous linking

[11] I have given instances in my work *On the Variation of Animals and Plants under Domestication*, ch. xv, 2nd edit., vol. ii, p. 69.

forms:[12] If it can be shown that the common wild oxlip, which is intermediate in character between the cowslip and primrose, resembles in sterility and other essential respects a hybrid plant, and if it can further be shown that the exlip, though in a high degree sterile, can be fertilized by either parent species, thus giving rise to still finer gradational links, then the presence of such linking forms in a state of nature ceases to be an argument of any weight in favour of the cowslip and primrose being varieties, and becomes, in fact, an argument on the other side. The hybrid origin of a plant in a state of nature can be recognized by four tests: first, by its occurrence only where both presumed parent / species exist or have recently existed; and this holds good, as far as I can discover, with the oxlip; but the *P. elatior* of Jacq., which, as we shall presently see, constitutes a distinct species, must not be confounded with the common oxlip. Secondly, by the supposed hybrid plant being nearly intermediate in character between the two parent species, and especially by its resembling hybrids artificially made between the same two species. Now the oxlip is intermediate in character, and resembles in every respect, except in the colour of the corolla, hybrids artificially produced between the primrose and the polyanthus, which latter is a variety of the cowslip. Thirdly, by the supposed hybrids being more or less sterile when crossed *inter se*: but to try this fairly, two distinct plants of the same parentage, and not two flowers on the same plant, should be crossed; for many pure species are more or less sterile with pollen from the same individual plant; and in the case of hybrids from heterostyled species the opposite forms should be crossed. Fourthly and lastly, by the supposed hybrids being much more fertile when crossed with either pure parent species than when crossed *inter se*, but still not as fully fertile as the parent species.

For the sake of ascertaining the two latter points, I transplanted a group of wild oxlips into my garden. They consisted of one long-styled and three short-styled plants, which, except in the corolla of one being slightly larger, resembled each other closely. The trials which were made, and the results obtained, are shown in the five following tables. No less than twenty different crosses are necessary in order to ascertain fully the fertility of hybrid heterostyled plants, both *inter se* and with their two parents species. In this instance 256 flowers / were crossed in the course of four seasons. I may mention, as a mere curiosity,

[12] See an excellent article on this subject by Mr H. C. Watson, in the *Phytologist*, vol. iii, p. 43.

that if anyone were to raise hybrids between two trimorphic hetero-styled species, he would have to make 90 distinct unions in order to ascertain their fertility in all ways; and as he would have to try at least 10 flowers in each case, he would be compelled to fertilize 900 flowers and count their seeds. This would probably exhaust the patience of the most patient man.

TABLE 14 *Crosses inter se between the two forms of the common oxlip*

Illegitimate union	Legitimate union	Illegitimate union	Legitimate union
Short-styled oxlip, by pollen of short-styled oxlip: 20 flowers fertilized, did not produce one capsule.	Short-styled oxlip, by pollen of long-styled oxlip: 10 flowers fertilized, did not produce one capsule.	Long-styled oxlip, by its own pollen: 24 flowers fertilized, produced five capsules, containing 6, 10, 20, 8, and 14 seeds. Average 11·6.	Long-styled oxlip, by pollen of short-styled oxlip: 10 flowers fertilized, did not produce one capsule.

TABLE 15 *Both forms of the oxlip crossed with pollen of both forms of the cowslip, P. veris*

Illegitimate union	Legitimate union	Illegitimate union	Legitimate union
Short-styled oxlip, by pollen of short-styled cowslip: 18 flowers fertilized, did not produce one capsule.	Short-styled oxlip, by pollen of long-styled cowslip: 18 flowers fertilized, produced three capsules, containing 7, 3, and 3 wretched seeds, apparently incapable of germination.	Long-styled oxlip, by pollen of long-styled cowslip: 11 flowers fertilized, produced one capsule, containing 13 wretched seeds.	Long-styled oxlip, by pollen of short-styled cowslip: 5 flowers fertilized, produced two capsules, containing 21 and 28 very fine seeds.

TABLE 16 *Both forms of the oxlip crossed with pollen of both forms of the primrose, P. vulgaris*

Illegitimate union	Legitimate union	Illegitimate union	Legitimate union
Short-styled oxlip, by pollen of short-styled primrose: 34 flowers fertilized, produced two capsules, containing 5 and 12 seeds.	Short-styled oxlip, by pollen of long-styled primrose: 26 flowers fertilized, produced six capsules, containing 16, 20, 5, 10, 19, and 24 seeds. Average 15·7. Many of the seeds very poor, some good.	Long-styled oxlip, by pollen of long-styled primrose: 11 flowers fertilized, produced four capsules, containing 10, 7, 5, and 6 wretched seeds. Average 7·0.	Long-styled oxlip, by pollen of short-styled primrose: 5 flowers fertilized, produced five capsules, containing 26, 32, 23, 28, and 34 seeds. Average 28·6.

47

TABLE 17   *Both forms of the cowslip crossed with pollen of both forms of the oxlip*

Illegitimate union	*Legitimate union*	*Illegitimate union*	*Legitimate union*
Short-styled cowslip, by pollen of short-styled oxlip: 8 flowers fertilized, produced not one capsule.	Long-styled cowslip, by pollen of short-styled oxlip: 8 flowers fertilized, produced one capsule, containing 26 seeds.	Long-styled cowslip, by pollen of long-styled oxlip: 8 flowers fertilized, produced three capsules, containing 5, 6, and 14 seeds. Average 8·3.	Short-styled cowslip, by pollen of long-styled oxlip: 8 flowers fertilized, produced eight capsules, containing 58, 38, 31, 44, 23, 26, 37, and 66 seeds. Average 40·4.

TABLE 18   *Both forms of the primrose crossed with pollen of both forms of the oxlip*

Illegitimate union	*Legitimate union*	*Illegitimate union*	*Legitimate union*
Short-styled primrose, by pollen of short-styled oxlip: 8 flowers fertilized, produced not one capsule.	Long-styled primrose, by pollen of short-styled oxlip: 8 flowers fertilized, produced two capsules, containing 5 and 2 seeds.	Long-styled primrose, by pollen of long-styled oxlip: 8 flowers fertilized, produced eight capsules, containing 15, 7, 12, 20, 22, 7, 16, and 13 seeds. Average 14·0.	Short-styled primrose, by pollen of long-styled oxlip: 8 flowers fertilized, produced four capsules, containing 52, 52, 42, and 49 seeds, some good and some bad. Average 48·7.

We see in these five tables the number of capsules and of seeds produced, by crossing both forms of the oxlip in a legitimate and illegitimate manner with one another, and with the two forms of the primrose and cowslip. I may premise that the pollen of two of the short-styled oxlips consisted of nothing but minute aborted whitish cells; but in the third short-styled plant about one-fifth of the grains appeared in a sound condition. Hence it is not surprising that neither the short-styled nor the long-styled oxlip produced a single seed when fertilized with this pollen. Nor did the pure cowslips or primroses when illegitimate fertilized with it; but when thus legitimately fertilized they yielded a few good seeds. The female organs of the short-styled oxlips, though greatly deteriorated in power, were in a rather better condition than the male organs; for though the short-styled oxlips yielded no seeds when fertilized by the long-styled oxlips, and hardly any when illegitimately fertilized by pure cowslips or primroses, yet when legitimately fertilized by these latter species, especially by the long-styled primrose, they yielded a moderate supply of good seed.

The long-styled oxlip was more fertile than the three short-styled oxlips, and about half its pollen grains appeared sound. It bore no seed when legitimately fertilized by the short-styled oxlips; but this no doubt was due to the badness of the pollen of the latter; for when illegitimately fertilized (Table 14) by its own pollen it produced some good seeds, though much fewer than self-fertilized cowslips or primroses would have produced. The long-styled oxlip likewise yielded a very low average of seed, as may be seen in the third compartment of the four latter tables, when illegitimately fertilized by, and when / illegitimately fertilizing, pure cowslips and primroses. The four corresponding legitimate unions, however, were moderately fertile, and one (viz. that between a short-styled cowslip and the long-styled oxlip in Table 17) was nearly as fertile as if both parents had been pure. A short-styled primrose legitimately fertilized by the long-styled oxlip (Table 18) also yielded a moderately good average, namely 48·7 seeds; but if this short-styled primrose had been fertilized by a long-styled primrose it would have yielded an average of 65 seeds. If we take the ten legitimate unions together, and the ten illegitimate unions together, we shall find that 29 per cent of the flowers fertilized in a legitimate manner yielded capsules, these containing on an average 27·4 good and bad seeds; whilst only 15 per cent of the flowers fertilized in an illegitimate manner yielded capsules, these containing on an average only 11·0 good and bad seeds.

In a previous part of this chapter it was shown that illegitimate crosses between the long-styled form of the primrose and the long-styled cowslip, and between the short-styled primrose and short-styled cowslip, are more sterile than legitimate crosses between these two species; and we now see that the same rule holds good almost invariably with their hybrid offspring, whether these are crossed *inter se*, or with either parent species; so that in this particular case, but not as we shall presently see in other cases, the same rule prevails with the pure unions between the two forms of the same heterostyled species, with crosses between two distinct heterostyled species, and with their hybrid offspring.

Seeds from the long-styled oxlip fertilized by its own pollen were sown, and three long-styled plants raised. The first of these was identical in every character with its parent. The second bore rather / smaller flowers, of a paler colour, almost like those of the primrose;

49

the scapes were at first single-flowered, but later in the season a tall thick scape, bearing many flowers, like that of the parent oxlip, was thrown up. The third plant likewise produced at first only single-flowered scapes, with the flowers rather small and of a darker yellow; but it perished early. The second plant also died in September; and the first plant, though all three grew under very favourable conditions, looked very sickly. Hence we may infer that seedlings from self-fertilized oxlips would hardly be able to exist in a state of nature. I was surprised to find that all the pollen grains in the first of these seedling oxlips appeared sound; and in the second only a moderate number were bad. These two plants, however, had not the power of producing a proper number of seeds; for though left uncovered and surrounded by pure primroses and cowslips, the capsules were estimated to include an average of only from fifteen to twenty seeds.

From having many experiments in hand, I did not sow the seed obtained by crossing both forms of the primrose and cowslip with both forms of the oxlip, which I now regret; but I ascertained an interesting point, namely, the character of the offspring from oxlips growing in a state of nature near both primroses and cowslips. The oxlips were the same plants which, after their seeds had been collected, were trans-planted and experimented on. From the seeds thus obtained eight plants were raised, which, when they flowered, might have been mistaken for pure primroses; but on close comparison the eye in the centre of the corolla was seen to be of a darker yellow and the peduncles more elongated. As the season advanced, one of these plants threw up two naked scapes, / 7 inches in height, which bore umbels of flowers of the same character as before. This fact led me to examine the other plants after they had flowered and were dug up; and I found that the flower-peduncles of all sprung from an extremely short common scape, of which no trace can be found in the pure primrose. Hence these plants are beautifully intermediate between the oxlip and the primrose, inclining rather towards the latter; and we may safely conclude that the parent oxlips had been fertilized by the surrounding primroses.

From the various facts now given, there can be no doubt that the common oxlip is a hybrid between the cowslip (*P. veris*, Brit. Fl.) and the primrose (*P. vulgaris*, Brit. Fl.), as has been surmised by several botanists. It is probable that oxlips may be produced either from the cowslip or the primrose and the seed-bearer, but oftenest from the latter, as I judge from the nature of the stations in which oxlips are generally found,[13] and from the primrose when crossed by the cowslip being more fertile than, conversely, the cowslip by the primrose. The hybrids themselves are also rather more fertile when crossed with the primrose than with the cowslip. Whichever may be the seed-bearing plant, the cross is probably between different forms of the two species; for we have seen that legitimate hybrid unions are more fertile than illegitimate hybrid unions. Moreover a friend in Surrey found that 29 oxlips which grew in the neighbourhood of his house consisted of 13 long-styled and 16 short-styled plants; now, if the parent plants had been illegitimately united, either the long- or short-styled form would have greatly preponderated, as we shall / hereafter see good reason to believe. The case of the oxlip is interesting; for hardly any other instance is known of a hybrid spontaneously arising in such large numbers over so wide an extent of country. The common oxlip (not the *P. elatior* of Jacq.) is found almost everywhere throughout England, where both cowslips and primroses grow. In some districts, as I have seen near Hartfield in Sussex and in parts of Surrey, specimens may be found on the borders of almost every field and small wood. In other districts the oxlip is comparatively rare: near my own residence I have found, during the last twenty-five years, not more than five or six plants or groups of plants. It is difficult to conjecture what is the cause of this difference in their number. It is almost necessary that a plant, or several plants, belonging to the same form, of one parent species, should grow near the opposite form of the other parent species; and it is further necessary that both species should be frequented by the same kind of insect, no doubt a moth. The cause of the rare appearance of the oxlip in certain districts may be the rarity of some moth, which in other districts habitually visits both the primrose and cowslip.

Finally, as the cowslip and primrose differ in the various characters above specified – as they are in a high degree sterile when intercrossed – as there is no trustworthy evidence that either species, when uncrossed, has ever given birth to the other species or to any intermediate form – and as the intermediate forms which are often found in a state of nature have been shown to be more or less sterile hybrids of the first or second generation – we must for the future look at the cowslip and primrose as good and true species. /

Primula elatior, Jacq., or the Bardfield Oxlip, is found in England

[13] See also on this head Hardwicke's *Science-Gossip*, 1867, pp. 114, 137.

only in two or three of the eastern counties. On the Continent it has a somewhat different range from that of the cowslip and primrose; and it inhabits some districts where neither of these species live.[14] In general appearance it differs so much from the common oxlip, that no one accustomed to see both forms in the living state could afterwards confound them; but there is scarcely more than a single character by which they can be distinctly defined, namely, their linear-oblong capsules equalling the calyx in length.[15] The capsules when mature differ conspicuously, owing to their length, from those of the cowslip and primrose. With respect to the fertility of the two forms when these are united in the four possible methods, they behave like the other heterostyled species of the genus, but differ somewhat (see Tables 8 and 12) in the smaller proportion of the illegitimately fertilized flowers which set capsules. That *P. elatior* is not a hybrid is certain, for when the two forms were legitimately united they yielded the large average of 47·1 seeds, and when illegitimately united 35·5 per capsule; whereas, of the four possible unions (Table 14) between the two forms of the common oxlip which we know to be a hybrid, one alone yielded any seed; and in this case the average number was only 11·6 per capsule. Moreover I could not detect a single bad pollen grain in the anthers of the short-styled *P. elatior*; whilst in two short-styled plants of the common oxlip all the grains were bad, as were a large majority in a third plant. As the common / oxlip is a hybrid between the primrose and cowslip, it is not surprising that eight long-styled flowers of the primrose, fertilized by pollen from the long-styled common oxlip, produced eight capsules (Table 18), containing, however, only a low average of seeds; whilst the same number of flowers of the primrose, similarly fertilized by the long-styled Bardfield oxlip, produced only a single capsule; this latter plant being an altogether distinct species from the primrose. Plants of *P. elatior* have been propogated by seed in a garden for twenty-five years, and have kept all this time quite constant, excepting that in some cases the flowers varied a little in size and tint.[16] Nevertheless, according to Mr H. C. Watson and Dr Bromfield,[17] plants may be occasionally found in a state of nature, in

[14] For England, see Hewett C. Watson, *Cybele Britannica*, vol. ii, 1849, p. 292. For the Continent, see Lecoq, *Géograph. Botanique de l'Europe*, vol. viii, 1858, p. 142. For the Alps, see *Ann. and Mag. Nat. Hist.*, vol. ix, 1842, pp. 156 and 515.

[15] Babington's *Manual of British Botany*, 1851, p. 258.

[16] See Mr H. Doubleday in the *Gardener's Chronicle*, 1867, p. 435, also Mr W. Marshall, ibid., p. 462.

[17] *Phytologist*, vol. i, p. 1001, and vol. iii, p. 695.

which most of the characters by which this species can be distinguished from *P. veris* and *vulgaris* fail; but such intermediate forms are probably due to hybridization; for Kerner states, in the paper before referred to, that hybrids sometimes, though rarely, arise in the Alps between *P. elatior* and *veris*.

Finally, although we may freely admit that *Primula veris, vulgaris*, and *elatior*, as well as all the other species of the genus, are descended from a common primordial form, yet from the facts above given, we must conclude that these three forms are now as fixed in character as are many others which are universally ranked as true species. Consequently they have as good a right to receive distinct specific names as have, for instance, the ass, quagga, and zebra.

Mr Scott has arrived at some interesting results by / crossing other heterostyled species of Primula.[18] I have already alluded to his statement, that in four instances (not to mention others) a species when crossed with a distinct one yielded a larger number of seeds than the same species fertilized illegitimately with its own-form pollen, though taken from a distinct plant. It has long been known from the researches of Kölreuter and Gärtner, that two species when crossed reciprocally sometimes differ as widely as is possible in their fertility: thus A when crossed with the pollen of B will yield a large number of seeds, whilst B may be crossed repeatedly with pollen of A, and will never yield a single seed. Now Mr Scott shows in several cases that the same law holds good when two heterostyled species of Primula are intercrossed, or when one is crossed with a homostyled species. But the results are much more complicated than with ordinary plants, as two heterostyled dimorphic species can be intercrossed in eight different ways. I will give one instance from Mr Scott. The long-styled *P. hirsuta* fertilized legitimately and illegitimately with pollen from the two forms of *P. auricula*, and reciprocally the long-styled *P. auricula* fertilized legitimately and illegitimately with pollen from the two forms of *P. hirsuta*, did not produce a single seed. Nor did the short-styled *P. hirsuta* when fertilized legitimately and illegitimately with the pollen of the two forms of *P. auricula*. On the other hand, the short-styled *P. auricula* fertilized with pollen from the long-styled *P. hirsuta* yielded capsules containing on an average no less than 56 seeds; and the short-styled *P. auricula* by pollen of the short-styled *P. hirsuta* yielded

[18] *Journ. Linn, Soc. Bot.*, vol. viii, 1864, p. 93 to end.

capsules containing on an average 42 seeds / per capsule. So that out of the eight possible unions between the two forms of these two speciés, six were utterly barren, and two fairly fertile. We have seen also the same sort of extraordinary irregularity in the results of my twenty different crosses (Tables 14 to 18), between the two forms of the oxlip, primrose, and cowslip. Mr Scott remarks, with respect to the results of his trials, that they are very surprising, as they show us that 'the sexual forms of a species manifest in their respective powers for conjunction with those of another species, physiological peculiarities which might well entitle them, by the criterion of fertility, to specific distinction'.

Finally, although *P. veris* and *vulgaris*, when crossed legitimately, and especially when their hybrid offspring are crossed in this manner with both parent species, were decidedly more fertile, than when crossed in an illegitimate manner, and although the legitimate cross effected by Mr Scott between *P. auricula* and *hirsuta* was more fertile, in the ratio of 56 to 42, than the illegitimate cross, nevertheless it is very doubtful, from the extreme irregularity of the results in the various other hybrid crosses made by Mr Scott, whether it can be predicted that two heterostyled species are generally more fertile if crossed legitimately (i.e. when opposite forms are united) than when crossed illegitimately.

Supplementary note on some wild hybrid verbascums

In an early part of this chapter I remarked that few other instances could be given of a hybrid spontaneously arising in such large numbers, and over so wide an extent of country, as that of the common oxlip; but perhaps the number of well-ascertained cases of naturally / produced hybrid willows is equally great.[19] Numerous spontaneous hybrids between several species of Cistus, found near Narbonne, have been carefully described by M. Timbal-Lagrave,[20] and many hybrids between an Aceras and Orchis have been observed by Dr Weddell.[21] In the genus Verbascum, hybrids are supposed to have often originated[22] in a state of nature; some of these unoubtedly are hybrids, and several hybrids have originated in gardens; but most of these cases require,[23] as Gärtner remarks, verification. Hence the following case is worth recording, more especially as the two species in question,

[19] Max Wichura, *Die Bastardbefruchtung, etc., der Weiden*, 1865.
[20] *Mém. de l'Acad. des Sciences de Toulouse*, 5th series, vol. v, p. 28.
[21] *Annales des Sc. Nat.*, 3rd series, *Bot.*, vol. xviii, p. 6.
[22] See, for instance, the *English Flora*, by Sir J. E. Smith, 1824, vol. i, p. 307.
[23] See Gärtner, *Bastarderzeugung*, 1849, p. 590.

V. thapsus and *lychnitis*, are perfectly fertile when insects are excluded, showing that the stigma, of each flower receives its own pollen. Moreover the flowers offer only pollen to insects, and have not been rendered attractive to them by secreting nectar.

I transplanted a young wild plant into my garden for experimental purposes, and when it flowered it plainly differed from the two species just mentioned and from a third which grows in this neighbourhood. I thought that it was a strange variety of *V. thapsus*. It attained the height (by measurement) of 8 feet! It was covered with a net, and ten flowers were fertilized with pollen from the same plant; later in the season, when uncovered, the flowers were freely visited by pollen-collecting bees; nevertheless, although many capsules were produced, not one contained a single seed. During the following year this same plant was / left uncovered near plants of *V. thapsus* and *lychnitis*; but again it did not produce a single seed. Four flowers, however, which were repeatedly fertilized with pollen of *V. lychnitis*, whilst the plant was temporarily kept under a net, produced four capsules, which contained five, one, two, and two seeds; at the same time three flowers were fertilized with pollen of *V. thapsus*, and these produced two, two, and three seeds. To show how unproductive these seven capsules were, I may state that a fine capsule from a plant of *V. thapsus*, growing close by, contained above 700 seeds. These facts led me to search the moderately sized field whence my plant had been removed, and I found in it many plants of *V. thapsus* and *lychnitis*, as well as thirty-three plants intermediate in character between these two species. These thirty-three plants differed much from one another. In the branching of the stem they more closely resembled *V. lychnitis* than *V. thapsus*, but in height the latter species. In the shape of their leaves they often closely approach *V. lychnitis*, but some had leaves extremely woolly on the upper surface and decurrent like those of *V. thapsus*; yet the degree of woolliness and of decurrency did not always go together. In the petals being flat and remaining open, and in the manner in which the anthers of the longer stamens were attached to the filaments, these plants all took more after *V. lychnitis* than *V. thapsus*. In the yellow colour of the corolla they all resembled the latter species. On the whole, these plants appeared to take rather more after *V. lychnitis* than *V. thapsus*. On the supposition that they were hybrids, it is not an anomalous circumstance that they should all have produced yellow flowers; for Gärtner crossed white and yellow-flowered varieties of Verbascum, and the offspring thus produced never bore flowers of an intermediate / tint, but either pure white or pure yellow flowers, generally of the latter colour.[24]

My observations were made in the autumn; so that I was able to collect some half-matured capsules from twenty of the thirty-three intermediate plants, and likewise capsules of the pure *V. lychnitis* and *thapsus* growing in the same field. All the latter were filled with perfect but immature seeds, whilst the capsules of the twenty intermediate plants did not contain one single perfect seed. These plants, consequently, were absolutely barren. From this fact – from the one plant which was transplanted into my garden yielding when artificially fertilized with pollen from *V. lychnitis* and *thapsus* some seeds, though extremely few in number – from the circumstances of the two pure species growing in the same field – and from the intermediate character of the sterile plants, there can be no doubt that they were hybrids. Judging from the position in which they were chiefly found, I am inclined to believe they were descended from *V. thapsus* as the seed-bearer, and *V. lychnitis* as the pollen-bearer.

It is known that many species of Verbascum, when the stem is jarred or struck by a stick, cast off their flowers.[25] This occurs with *V. thapsus*, as I have repeatedly observed. The corolla first separates from its attachment, and then the sepals spontaneously bend inwards so as to clasp the ovarium, pushing off the corolla by their movement, in the course of two or three minutes. Nothing of this kind takes place with young barely expanded flowers. With *Verbascum lychnitis* and, as I believe, *V. phoeniceum* the corolla is not cast / off, however often and severely the stem may be struck. In this curious property the above-described hybrids took after *V. thapsus*; for I observed, to my surprise, that when I pulled off the flower-buds round the flowers which I wished to mark with a thread, the slight jar invariably caused the corollas to fall off.

These hybrids are interesting under several points of view. First, from the number found in various parts of the same moderately-sized field. That they owed their origin to insects flying from flower to flower, whilst collecting pollen, there can be no doubt. Although insects thus rob the flowers of a most precious substance, yet they do

---

[24] *Bastarderzeugung*, p. 307.

[25] This was first oberved by Correa de Serra: see Sir J. E. Smith's *English Flora*, 1824, vol. i, p. 311; also *Life of Sir J. E. Smith*, vol. ii, p. 210. I was guided to these references by the Rev. W. A. Leighton, who observed this same phenomenon with *V. virgatum*.

great good; for as I have elsewhere shown,[26] the seedlings of V. *thapsus* raised from flowers fertilized with pollen from another plant, are more vigorous than those raised from self-fertilized flowers. But in this particular instance the insects did great harm, as they led to the production of utterly barren plants. Secondly, these hybrids are remarkable from differing much from one another in many of their characters; for hybrids of the first generation, if raised from uncultivated plants, are generally uniform in character. That these hybrids belonged to the first generation we may safely conclude, from the absolute sterility of all those observed by me in a state of nature and of the one plant in my garden, excepting when artificially and repeatedly fertilized with pure pollen, and then the number of seeds produced was extremely small. As these hybrids varied so much, an almost perfectly graduated series of forms, connecting together the two widely distinct parent species, could easily have been selected. This case, like that of the common oxlip, shows that botanists ought to be / cautious in inferring the specific identity of two forms from the presence of intermediate gradations; nor would it be easy in the many cases in which hybrids are moderately fertile to detect a slight degree of sterility in such plants growing in a state of nature and liable to be fertilized by either parent species. Thirdly and lastly, these hybrids offer an excellent illustration of a statement made by that admirable observer Gärtner, namely, that although plants which can be crossed with ease generally produce fairly fertile offspring, yet well-pronounced exceptions to this rule occur; and here we have two species of Verbascum which evidently cross with the greatest ease, but produce hybrids which are excessively sterile. /

[26] *The Effects of Cross and Self-fertilisation,* 1876, p. 89.

CHAPTER III

HETEROSTYLED DIMORPHIC PLANTS – *continued*

> *Linum grandiflorum*, long-styled form utterly sterile with own-form pollen
> – *Linum perenne*, torsion of the pistils in the long-styled form alone –
> Homostyled species of Linum – *Pulmonaria officinalis*, singular difference
> in self-fertility between the English and German long-styled plants –
> *Pulmonaria angustifolia* shown to be a distinct species, long-styled form
> completely self-sterile – *Polygonum fagopyrum* – Various other heterostyled
> genera – Rubiaceae – *Mitchella repens*, fertility of the flowers in pairs –
> Houstonia – Faramea, remarkable difference in the pollen-grains of the
> two forms; torsion of the stamens in the short-styled form alone;
> development not as yet perfect – The heterostyled structure in the several
> Rubiaceous genera not due to descent in common.

It has long been known[1] that several species of Linum present two forms, and having observed this fact in *L. flavum* more than thirty years ago, I was led, after ascertaining the nature of heterostylism in Primula, to examine the first species of Linum which I met with, namely, the beautiful *L. grandiflorum*. This plant exists under two forms, occurring in about equal numbers, which differ little in structure, but greatly in function. The foliage, corolla, stamens, and pollen grains (the latter examined both distended with water and dry) are alike in the two forms (Fig. 4). The difference is confined to the pistil; in the short-styled form the styles and the stigmas are only about half the length of those in the long-styled. A more / important distinction is, that the five stigmas in the short-styled form diverge greatly from one another, and pass out between the filaments of the stamens, and thus lie within the tube of the corolla. In the long-styled form the elongated stigmas stand nearly upright, and alternate with the anthers. In this latter form the length of the stigmas varies considerably, their upper extremities projecting even a little above the anthers, or reaching up only to about their middle. Nevertheless, there

---

[1] Treviranus has shown that this is the case in his review of my original paper, *Bot. Zeitung*, 1863, p. 189.

Long-styled form   Short-styled form
Fig. 4
s s stigmas
Linum grandiflorum

is never the slightest difficulty in distinguishing between the two forms; for, besides the difference in the divergence of the stigmas, those of the short-styled form never reach even to the bases of the anthers. In this form the papillae on the stigmatic surfaces are shorter, darker coloured, and more crowded together than in the long-styled form; but these differences seem due merely to the shortening of the stigma, for in the varieties of the long-styled form with shorter stigmas, the papillae are more crowded and darker-coloured than in those with the longer / stigmas. Considering the slight and variable differences between the two forms of this Linum, it is not surprising that hitherto they have been overlooked.

In 1861 I had eleven plants in my garden, eight of which were long-styled, and three short-styled. Two very fine long-styled plants grew in a bed a hundred yards off all the others, and separated from them by a screen of evergreens. I marked twelve flowers, and placed on their stigmas a little pollen from the short-styled plants. The pollen of the two forms is, as stated, identical in appearance; the stigmas of the long-styled flowers were already thickly covered with their own pollen – so thickly that I could not find one bare stigma, and it was late in the season, namely, 15 September. Altogether, it seemed almost childish to expect any result. Nevertheless from my experiments on Primula, I had faith, and did not hesitate to make the trial, but certainly did not anticipate the full result which was obtained. The germens of these twelve flowers all swelled, and ultimately six fine capsules (the seed of

59

which germinated on the following year) and two poor capsules were produced; only four capsules shanking off. These same two long-styled plants produced, in the course of the summer, a vast number of flowers, the stigmas of which were covered with their own pollen; but they all proved absolutely barren, and their germens did not even swell.

The nine other plants, six long-styled and three short-styled, grew not very far apart in my flower-garden. Four of these long-styled plants produced no seed-capsules; the fifth produced two; and the remaining one grew so close to a short-styled plant that their branches touched, and this produced twelve capsules, but they were poor ones. The case was different / with the short-styled plants. The one which grew close to the long-styled plant produced ninety-four imperfectly fertilized capsules containing a multitude of bad seeds, with a moderate number of good ones. The two other short-styled plants growing together were small, being partly smothered by other plants; they did not stand very close to any long-styled plants, yet they yielded together nineteen capsules. These facts seem to show that the short-styled plants are more fertile with their own pollen than are the long-styled, and we shall immediately see that this probably is the case. But I suspect that the difference in fertility between the the two forms was in this instance in part due to a distinct cause. I repeatedly watched the flowers, and only once saw a humble-bee momentarily alight on one, and then fly away. If bees had visited the several plants, there cannot be a doubt that the four long-styled plants, which did not produce a single capsule, would have borne an abundance. But several times I saw small diptera sucking the flowers; and these insects, though not visiting the flowers with anything like the regularity of bees, would carry a little pollen from one form to the other, especially when growing near together; and the stigmas of the short-styled plants, diverging within the tube of the corolla, would be more likely than the upright stigmas of the long-styled plants, to receive a small quantity of pollen if brought to them by small insects. Moreover from the greater number of the long-styled than of the short-styled plants in the garden, the latter would be more likely to receive pollen from the long-styled, than the long-styled from the short-styled.

In 1862 I raised thirty-four plants of this Linum in a hot-bed; and these consisted of seventeen long-styled and seventeen short-styled forms. Seed sown later in the / flower-garden yielded seventeen long-styled and twelve short-styled forms. These facts justify the statement that the two forms are produced in about equal numbers. The thirty-four plants of the first lot were kept under a net which excluded all insects, except such minute ones as thrips. I fertilized fourteen long-styled flowers legitimately, with pollen from the short-styled, and got eleven fine seed-capsules, which contained on an average 8·6 seeds per capsule, but only 5·6 appeared to be good. It may be well to state that ten seeds is the maximum production for a capsule, and that our climate cannot be very favourable to this North-African plant. On three occasions the stigmas of nearly a hundred flowers were fertilized illegitimately with their own-form pollen, taken from separate plants, so as to prevent any possible ill effects from close inter-breeding. Many other flowers were also produced, which, as before stated, must have received plenty of their own pollen; yet from all these flowers, borne by the seventeen long-styled plants, only three capsules were produced. One of these included no seed, and the other two together gave only five good seeds. It is probable that this miserable product of two half-fertile capsules from the seventeen plants, each of which must have produced at least fifty or sixty flowers, resulted from their fertilization with pollen from the short-styled plants by the aid of thrips; for I made a great mistake in keeping the two forms under the same net, with their branches often interlocking; and it is surprising that a greater number of flowers were not accidentally fertilized.

Twelve short-styled flowers were in this instance castrated, and afterwards fertilized legitimately with pollen from the long-styled form; and they produced seven fine capsules. These included on an average / 7·6 seeds, but of apparently good seed only 4·3 per capsule. At three separate times nearly a hundred flowers were fertilized illegitimately with their own-form pollen, taken from separate plants; and numerous other flowers were produced, many of which must have received their own pollen. From all these flowers on the seventeen short-styled plants only fifteen capsules were produced, of which only eleven contained any good seed, on an average 4·2 per capsule. As remarked in the case of the long-styled plants, some even of these capsules were perhaps the product of a little pollen accidentally fallen from the adjoining flowers of the other form on to the stigmas, or transported by thrips. Nevertheless the short-styled plants seem to be slightly more fertile with their own pollen than the long-styled, in the proportion of fifteen capsules to three; nor can this difference be accounted for by the short-styled stigmas being more liable to receive their own pollen than the long-styled, for the reverse is the case. The greater self-fertility of the short-styled flowers was likewise shown in 1861 by the plants in my flower-garden, which were left to themselves, and were but sparingly visited by insects.

On account of the probability of some of the flowers on the plants of both forms, which were covered under the same net, having been legitimately fertilized in an accidental manner, the relative fertility of the two legitimate and two illegitimate unions cannot be compared with certainty; but judging from the number of good seeds per capsule, the difference was at least in the ratio of 100 to 7, and probably much greater.

Hildebrand tested my results, but only on a single short-styled plant, by fertilizing many flowers with their own-form pollen; and these did not produce any / seed. This confirms my suspicion that some of the few capsules produced by the foregoing seventeen short-styled plants were the product of accidental legitimate fertilization. Other flowers on the same plant were fertilized by Hildebrand with pollen from the long-styled form, and all produced fruit.[2]

The absolute sterility (judging from the experiments of 1861) of the long-styled plants with their own-form pollen led me to examine into its apparent cause; and the results are so curious that they are worth giving in detail. The experiments were tried on plants grown in pots and brought successively into the house.

First. Pollen from a short-styled plant was placed on the five stigmas of a long-styled flower, and these, after thirty hours, were found deeply penetrated by a multitude of pollen tubes, far too numerous to be counted; the stigmas had also become discoloured and twisted. I repeated this experiment on another flower, and in eighteen hours the stigmas were penetrated by a multitude of long pollen tubes. This is what might have been expected, as the union is a legitimate one. The converse experiment was likewise tried, and pollen from a long-styled flower was placed on the stigmas of a short-styled flower, and in twenty-four hours the stigmas were discoloured, twisted, and penetrated by numerous pollen tubes; and this, again, is what might have been expected, as the union was a legitimate one.

Secondly. Pollen from a long-styled flower was placed on all five stigmas of a long-styled flower on a separate plant: after nineteen hours the stigmas were dissected, and only a single pollen grain had emitted a tube, / and this was a very short one. To make sure that the

[2] *Bot. Zeitung*, 1 January, 1864, p. 2.

pollen was good, I took in this case, and in most of the other cases, pollen either from the same anther or from the same flower, and proved it to be good by placing it on the stigma of a short-styled plant, and found numerous pollen tubes emitted.

Thirdly. Repeated last experiment, and placed own-form pollen on all five stigmas of a long-styled flower; after nineteen hours and a half, not one single grain had emitted its tube.

Fourthly. Repeated the experiment, with the same result after twenty-four hours.

Fifthly. Repeated last experiment, and, after leaving pollen on for nineteen hours, put on an additional quantity of own-form pollen on all five stigmas. After an interval of three days, the stigmas were examined, and, instead of being discoloured and twisted, they were straight and fresh-coloured. Only one grain had emitted a quite short tube, which was drawn out of the stigmatic tissue without being ruptured.

The following experiments are more striking:

Sixthly. I placed own-form pollen on three of the stigmas of the long-styled flower, and pollen from a short-styled flower on the other two stigmas. After twenty-two hours these two stigmas were discoloured, slightly twisted, and penetrated by the tubes of numerous pollen grains: the other three stigmas, covered with their own-form pollen, were fresh, and all the pollen grains were loose; but I did not dissect the whole stigma.

Seventhly. Experiment repeated in the same manner, with the same result.

Eighthly. Experiment repeated, but the stigmas were carefully examined after an interval of only five hours and a half. The two stigmas with pollen from a / short-styled flower were penetrated by innumerable tubes, which were as yet short, and the stigmas themselves were not at all discoloured. The three stigmas covered with their own-form pollen were not penetrated by a single pollen tube.

Ninthly. Put pollen of a short-styled flower on a single long-styled stigma, and own-form pollen on the other four stigmas; after twenty-four hours the one stigma was somewhat discoloured and twisted, and penetrated by many long tubes: the other four stigmas were quite straight and fresh; but on dissecting them I found that three pollen grains had protruded very short tubes into the tissue.

Tenthly. Repeated the experiment, with the same result after twenty-four hours, excepting that only two own-form grains had penetrated the stigmatic tissue with their tubes to a very short depth. The one stigma, which was deeply penetrated by a multitude of tubes from the short-styled pollen, presented a conspicuous difference in being much curled, half-shrivelled, and discoloured, in comparison with the other four straight and bright pink stigmas.

I could add other experiments: but those now given amply suffice to show that the pollen grains of a short-styled flower placed on the stigma of a long-styled flower emit a multitude of tubes after an interval of from five to six hours, and penetrate the tissue ultimately to a great depth; and that after twenty-four hours the stigmas thus penetrated change colour, become twisted, and appear half-withered. On the other hand, pollen grains from a long-styled flower placed on its own stigmas, do not emit their tubes after an interval of a day, or even three days; or at most only three or four grains out of a multitude emit their tubes, and these apparently never penetrate the / stigmatic tissue deeply, and the stigmas themselves do not soon become discoloured and twisted.

This seems to me a remarkable physiological fact. The pollen grains of the two forms are undistinguishable under the microscope; the stigmas differ only in length, degree of divergence, and in the size, shade of colour, and approximation of their papillae, these latter differences being variable and apparently due merely to the degree of elongation of the stigma. Yet we plainly see that the two kinds of pollen and the two stigmas are widely dissimilar in their mutual reaction – the stigmas of each form being almost powerless on their own pollen, but causing, through some mysterious influence, apparently by simple contact (for I could detect no viscid secretion), the pollen grains of the opposite form to protrude their tubes. It may be said that the two pollens and the two stigmas mutually recognize each other by some means. Taking fertility as the criterion of distinctness, it is no exaggeration to say that the pollen of the long-styled *Linum grandiflorum* (and conversely that of the other form) has been brought to a degree of differentiation, with respect to its action on the stigma of the same form, corresponding with that existing between the pollen and stigma of species belonging to distinct genera.

Linum perenne. This species is conspicuously heterostyled, as has been noticed by several authors. The pistil in the long-styled form is nearly twice as long as that of the short-styled. In the latter the stigmas are smaller and, diverging to a greater degree, pass out low down between the filaments. I could detect no difference in the two forms in the size of the stigmatic papillae. In the long-styled form alone the stigmatic surfaces of the mature pistils twist round, so as to face the circumference of the flower; but to this point I / shall presently return. Differently from what occurs in *L. grandiflorum*, the long-styled flowers have stamens hardly more than half the length of those in the short-styled. The size of the pollen grains is rather variable; after some doubt, I have come to the conclusion that there is no uniform difference between the grains in the two forms. The long stamens in the short-styled form project to some height above the corolla, and their filaments are coloured blue apparently from exposure to the light. The anthers of the longer stamens correspond in height with the lower part of the stigmas of the long-styled flowers; and the anthers of the shorter stamens of the latter correspond in the same manner in height with the stigmas of the short-styled flowers.

I raised from seed twenty-six plants, of which twelve proved to be long-styled and fourteen short-styled. They flowered well, but were not large plants. As I did not expect them to flower so soon, I did not transplant them, and they unfortunately grew with their branches closely interlocked. All the plants were covered under the same net, excepting one of each form. Of the flowers on the long-styled plants, twelve were illegitimately fertilized with their own-form pollen, taken in every case from a separate plant; and not one set a seed-capsule: twelve other flowers were legitimately fertilized with pollen from short-styled flowers; and they set nine capsules, each including on an average 7 good seeds, ten being the maximum number ever produced. Of the flowers on the short-styled plants, twelve were illegitimately fertilized with own-form pollen, and they yielded one capsule, including only 3 good seeds; twelve other flowers were legitimately fertilized with pollen from long-styled flowers, and these produced nine capsules, but one was bad; / the eight good capsules contained on an average 8 good seeds each. Judging from the number of seeds per capsule, the fertility of the two legitimate to that of the two illegitimate unions is as 100 to 20.

The numerous flowers on the eleven long-styled plants under the net, which were not fertilized, produced only three capsules, including 8, 4, and 1 good seeds. Whether these three capsules were the product of accidental legitimate fertilization, owing to the branches of the plants of the two forms interlocking, I will not pretend to decide. The single long-styled plant which was left uncovered, and grew close by the uncovered short-styled plant, produced five good pods; but it was a poor and small plant.

The flowers borne on the thirteen short-styled plants under the net, which were not fertilized, produced twelve capsules, containing on an average 5·6 seeds. As some of these capsules were very fine, and as five were borne on one twig, I suspect that some minute insect had accidentally got under the net and had brought pollen from the other form to the flowers which produced this little group of capsules. The one uncovered short-styled plant which grew close to the uncovered long-styled plant yielded twelve capsules.

From these facts we have some reason to believe, as in the case of *L. grandiflorum*, that the short-styled plants are in a slight degree more fertile with their own pollen than are the long-styled plants. Anyhow we have the clearest evidence, that the stigmas of each form require for full fertility that pollen from the stamens of corresponding height belonging to the opposite form should be brought to them.

Hildebrand, in the paper lately referred to, confirms my results. He placed a short-styled plant in his house, and fertilized about 20 flowers with their own / pollen, and abotut 30 with pollen from another plant belonging to the same form, and these 50 flowers did not set a single capsule. On the other hand he fertilized about 30 flowers with pollen from the long-styled form, and these, with the exception of two, yielded capsules, containing good seeds.

It is a singular fact, in contrast with what occurred in the case of *L. grandiflorum*, that the pollen grains of both forms of *L. perenne*, when placed on their own-form stigmas, emitted their tubes, though this action did not lead to the production of seeds. After an interval of eighteen hours, the tubes penetrated the stigmatic tissue, but to what depth I did not ascertain. In this case the impotence of the pollen grains on their own stigmas must have been due either to the tubes not reaching the ovules, or to their not acting properly after reaching them.

The plants both of *L. perenne* and *grandiflorum* grew, as already stated, with their branches interlocked, and with scores of flowers of the two forms close together; they were covered by a rather coarse net, through which the wind, when high, passed; and such minute insects as thrips could not, of course, be excluded; yet we have seen that the utmost possible amount of accidental fertilization on seventeen long-styled plants in the one case, and on eleven long-styled plants in the other, resulted in the production, in each case, of three poor capsules; so that when the proper insects are excluded, the wind does hardly anything in the way of carrying pollen from plant to plant. I allude to

THE DIFFERENT FORMS OF FLOWERS

this fact because botanists, in speaking of the fertilization of various flowers, often refer to the wind or to insects as if the alternative were indifferent. This view, according to my experience, is entirely erroneous. When the wind is the agent in carrying pollen, either from / one sex to the other, or from hermaphrodite to hermaphrodite, we can recognize structure as manifestly adapted to its action as to that of insects when these are the carriers. We see adaptation to the wind in the incoherence of the pollen – in the inordinate quantity produced (as in the Coniferae, Spinage, etc.) – in the dangling anthers well fitted to shake out the pollen – in the absence or small size of the perianth – in the protrusion of the stigmas at the period of fertilization – in the flowers being produced before they are hidden by the leaves – and in the stigmas being downy or plumose (as in the Gramineae, Docks, etc.), so as to secure the chance-blown grains. In plants which are fertilized by the wind, the flowers do not secrete nectar, their pollen is too incoherent to be easily collected by insects, they have not bright-coloured corollas to serve as guides, and they are not, as far as I have seen, visited by insects. When insects are the agents of fertilization (and this is incomparably the more frequent case with hermaphrodite plants), the wind plays no part, but we see an endless number of adaptations to ensure the safe transport of the pollen by the living workers. These adaptations are most easily recognized in irregular flowers: but they are present in regular flowers, of which those of Linum offer a good instance, as I will now endeavour to show.

I have already alluded to the rotation of each separate stigma in the long-styled form of *Linum perenne*. In both forms of the other heterostyled species and in the homostyled species of Linum which I have seen, the stigmatic surfaces face the centre of the flower, with the furrowed backs of the stigmas, to which the styles are attached, facing outwards. This is the case with the stigmas of the long-styled flowers of *L. perenne* whilst in bud. But by the time the flowers / have expanded, the five stigmas twist round so as to face the circumference, owing to the torsion of that part of the style which lies beneath the stigma. I should state that the five stigmas do not always turn round completely, two or three sometimes facing only obliquely outwards. My observations were made during October; and it is not improbable that earlier in the season the torsion would have been more complete; for after two or three cold and wet days the movement was very imperfectly performed. The flowers should be examined shortly after their expansion, as their duration is brief; as soon as they begin to wither, the styles become spirally twisted all together, the original position of the parts being thus lost.

He who will compare the structure of the whole flower in both forms of *L. perenne* and *grandiflorum*, and, as I may add, of *L. flavum*, will not doubt about the meaning of this torsion of the styles in the one form alone of *L. perenne*, as well as the meaning of the divergence of the stigmas in the short-styled form of all three species. It is absolutely necessary, as we know, that insects should carry pollen from the flowers of the one form reciprocally to those of the other. Insects are attracted by five drops of nectar, secreted exteriorly at the base of the stamens, so that to reach these drops they must insert their proboscides outside the ring of broad filaments, between them and the petals. In the short-styled form of the above three species, the stigmas face the axis of the flower; and had the styles retained their original upright and central position, not only would the stigmas have presented their backs to the insects which sucked the flowers, but their front and fertile surfaces would have been separated from the entering insects / by the ring of broad filaments, and would never have received any pollen. As it is, the styles diverge and pass out between the filaments. After this movement the short stigmas lie within the tube of the corolla; and their papillous surfaces being now turned upwards are necessarily brushed by every entering insect, and thus receive the required pollen.

In the long-styled form of *L. grandiflorum*, the almost parallel or slightly diverging anthers and stigmas project a little above the tube of the somewhat concave flower; and they stand directly over the open space leading to the drops of nectar. Consequently when insects visit the flowers of either form (for the stamens in this species occupy the same position in both forms), they will get their foreheads or proboscides well dusted with the coherent pollen. As soon as they visit the flowers of the long-styled form they will necessarily leave pollen on the proper surface of the elongated stigmas; and when they visit the short-styled flowers, they will leave pollen on the upturned stigmatic surfaces. Thus the stigmas of both forms will receive indifferently the pollen of both forms; but we know that the pollen alone of the opposite form causes fertilization.

In the case of *L. perenne*, affairs are arranged more perfectly; for the stamens in the two forms stand at different heights, so that pollen from the anthers of the longer stamens will adhere to one part of an insect's body, and will afterwards be brushed off by the rough stigmas of the longer pistils; whilst pollen from the anthers of the shorter stamens will adhere to a different part of the insect's body, and will afterwards be brushed off by the stigmas of the shorter pistils; and this is what is required for the legitimate fertilization of both forms. The corolla of *L. perenne* is more / expanded than that of *L. grandiflorum*, and the stigmas of the long-styled form do not diverge greatly from one another; nor do the stamens of either form. Hence insects, especially rather small ones, will not insert their proboscides between the stigmas of the long-styled form, nor between the anthers of either form (Fig. 5), but will strike against them, at nearly right angles, with the backs of their head or thorax. Now, in the long-styled flowers, if each stigma did not rotate on its axis, insects in visiting them would strike their heads against the backs of the stigmas; as it is, they strike against that surface which is covered / with papillae, with their heads already charged with pollen from the stamens of corresponding height borne by the flowers of the other form, and legitimate fertilization is thus ensured.

Thus we can understand the meaning of the torsion of the styles in

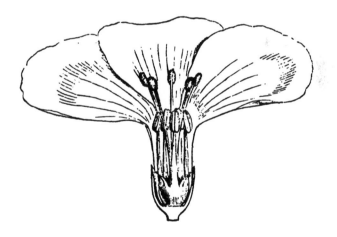

Fig. 5   Long-styled form of *L. perenne*, var. *Austriacum* in its early condition before the stigmas have rotated. The petals and calyx have been removed on the near side.[3]

[3] I neglected to get drawings made from fresh flowers of the two forms. But Mr Fitch has made the above sketch of a long-styled flower from dried specimens and from published engravings. His well-known skill ensures accuracy in the proportional size of the parts.

the long-styled flowers alone, as well as their divergence in the short-styled flowers.

One other point is worth notice. In botanical works many flowers are said to be fertilized in the bud. This statement generally rests, as far as I can discover, on the anthers opening in the bud; no evidence being adduced that the stigma is at this period mature, or that it is not subsequently acted on by pollen brought from other flowers. In the case of *Cephalanthera grandiflora* I have shown[4] that precocious and partial self-fertilization, with subsequent full fertilization, is the regular course of events. The belief that the flowers of many plants are fertilized in the bud, that is, are perpetually self-fertilized, is a most effectual bar to understanding their real structure. I am, however, far from wishing to assert that some flowers, during certain seasons, are not fertilized in the bud; for I have reason to believe that this is the case. A good observer,[5] resting his belief on the usual kind of evidence, states that in *Linum Austriacum* (which is heterostyled, and is considered by Planchon as a variety of *L. perenne*) the anthers open the evening before the expansion of the flowers, and that the stigmas are then almost always fertilized. Now we knwow positively that, so far from *Linum perenne* being fertilized by its own pollen in the bud, its own pollen is as powerless on the stigma as so much inorganic dust.

Linum flavum. The pistil of the long-styled form / of this species is nearly twice as long as that of the short-styled; the stigmas are longer and the papillae coarser. In the short-styled form the stigmas diverge and pass out between the filaments, as in the previous species. The stamens in the two forms differ in length; and, what is singular, the anthers of the longer stamens are not so long as those of the other form; so that in the short-styled form both the stigmas and the anthers are shorter than in the long-styled form. The pollen grains of the two forms do not differ in size. As this species is propagated by cuttings, generally all the plants in the same garden belong to the same form. I have enquired, but have never heard of its seeding in this country. Certainly my own plants never produced a single seed as long as I possessed only one of the two forms. After considerable search I procured both forms, but from want of time only a few experiments were made. Two plants of the two forms were planted some way apart

---

[4] *Fertilisation of Orchids*, p. 108 – 2nd edit., 1877, p. 84.
[5] *Etudes sur la Géogr. Bot.*, H. Lecoq, 1856, vol. v, p. 325.

in my garden, and were not covered by nets. Three flowers on the long-styled plant were legitimately fertilized with pollen from the short-styled plant, and one of them set a fine capsule. No other capsules were produced by this plant. Three flowers on the short-styled plant were legitimately fertilized with pollen from the long-styled, and all three produced capsules, containing respectively no less than 8, 9, and 10 seeds. Three other flowers on this plant, which had not been artificially fertilized, produced capsules containing 5, 1, and 5 seeds; and it is quite possible that pollen may have been brought to them by insects from the long-styled plant growing in the same garden. Nevertheless as they did not yield half the number of seeds compared with the other flowers on the same plant which had been artificially and legitimately fertilized, and as the / short-styled plants of the two previous species apparently evince some slight capacity for fertilization with their own-form pollen, these three capsules may have been the product of self-fertilization.

Besides the three species now described, the yellow-flowered *L. corymbiferum* is certainly heterostyled, as is, according to Planchon,[6] *L. salsoloides*. This botanist is the only one who seems to have inferred that heterostylism might have some important functional bearing. Dr Alefeld, who has made a special study of the genus, says[7] that about half of the sixty-five species known to him are heterostyled. This is the case with *L. trigynum*, which differs so much from the other species that it has been formed by him into a distinct genus.[8] According to the same author, none of the species which inhabit America and the Cape of Good Hope are heterostyled.

I have examined only three homostyled species, namely, *L. usitatissimum, angustifolium*, and *catharticum*. I raised 111 plants of a variety of the first-named species, and these, when protected under a net, all produced plenty of seed. The flowers, according to H. Müller,[9] are frequented by bees and moths. With respect to *L. catharticum*, the same author shows that the flowers are so constructed that they can freely fertilize themselves; but if visited by insects they might be cross-fertilized. He has, however, only once seen the flowers thus visited

[6] Hooker's *London Journal of Botany*, 1848, vol. vii, p. 174.

[7] *Bot. Zeitung*, 18 September, 1863, p. 281.

[8] It is not improbable that the allied genus, Hugonia, is heterostyled, for one species is said by Planchon (Hooker's *London Journal of Botany*, 1848, vol. vii, p. 525) to be provided with 'staminibus exsertis'; another with 'stylis staminibus longioribus', and another has 'stamina 5, majora, stylos longe superantia'.

[9] *Die Befruchtung der Blumen*, etc., p. 168.

during the day; but it / may be suspected that they are frequented during the night by small moths for the sake of the five minute drops of nectar secreted. Lastly, *L. Lewisii* is said by Planchon to bear on the same plant flowers with stamens and pistils of the same height, and others with the pistils either longer or shorter than the stamens. This case formerly appeared to me an extraordinary one; but I am now inclined to believe that it is one merely of great variability.[10]

PULMONARIA (BORAGINEAE)

Pulmonaria officinalis. Hildebrand has published[11] a full account of this heterostyled plant. The pistil of the long-styled form is twice as long as that of the short-styled; and the stamens differ in a corresponding, though converse, manner. There is no marked difference in the shape or state of surface of the stigma in the two forms. The pollen grains of the short-styled form are to those of the long-styled as 9 to 7, or as 100 to 78, in length, and as 7 to 6 in breadth. They do not differ in the appearance of their contents. The corolla of the one form differs in shape from that of the other in nearly the same manner as in Primula; but besides this difference the flowers of the short-styled are generally the larger of the two. Hildebrand collected on the Siebengebirge, ten wild long-styled and ten short-styled plants. The former bore 289 flowers, of which 186 (i.e. 64 per cent) had set fruit, yielding 1·88 seed per fruit. The ten short-styled plants bore 373 flowers, of which 262 (i.e. / 70 per cent) had set fruit, yielding 1·86 seed per fruit. So that the short-styled plants produced many more flowers, and these set a rather larger proportion of fruit, but the fruits themselves yielded a slightly lower average number of seeds than did the long-styled plants. The results of Hildebrand's experiments on the fertility of the two forms are given in Table 19.

In the summer of 1864, before I had heard of Hildebrand's experiments, I noticed some long-styled plants of this species (named for me by Dr Hooker) growing by themselves in a garden in Surrey; and to my surprise about half the flowers had set fruit, several of which contained 2, and one contained even 3 seeds. These seeds were sown

[10] Planchon, in Hooker's *London Journal of Botany*, 1848, vol. vii, p. 175. See on this subject Asa Gray, in *American Journal of Science*, vol. xxxvi, September, 1863, p. 284.
[11] *Bot. Zeitung*, 1865, 13 January, p. 13.

TABLE 19 *Pulmonaria officinalis (from Hildebrand)*

Nature of union	Number of flowers fertilized	Number of fruits produced	Average number of seeds per fruit
Long-styled flowers, by pollen of short-styled. Legitimate union	14	10	1·30
Long-styled flowers, 14 by own-pollen, and 16 by pollen of other plant of same form. Illegitimate union	30	0	0
Short-styled flowers by pollen of long-styled. Legitimate union	16	14	1·57
Short-styled flowers, 11 by own pollen, 14 by pollen of other plant of same form. Illegitimate union	25	0	0

in my garden and eleven seedlings thus raised, all of which proved long-styled, in accordance with the usual rule in such cases. Two years afterwards the plants were left uncovered, no / other plant of the same genus growing in my garden, and the flowers were visited by many bees. They set an abundance of seeds: for instance, I gathered from a single plant rather less than half of the seeds which it had produced, and they numbered 47. Therefore this illegitimately fertilized plant must have produced about 100 seeds; that is, thrice as many as one of the wild long-styled plants collected on the Siebengebirge by Hildebrand, and which, no doubt, had been legitimately fertilized. In the following year one of my plants was covered by a net, and even under these unfavourable conditions it produced spontaneously a few seeds. It should be observed that as the flowers stand either almost horizontally or hang considerably downwards, pollen from the short stamens would be likely to fall on the stigma. We thus see that the English long-styled plants when illegitimately fertilized were highly fertile, whilst the German plants similarly treated by Hildebrand were completely sterile. How to account for this wide discordance in our results I know not. Hildebrand cultivated his plants in pots and kept them for a time in the house, whilst mine were grown out of doors; and he thinks that this difference of treatment may have caused the difference in our results. But this does not appear to me nearly a sufficient cause, although his plants were slightly less productive than the wild ones growing on the Siebengebirge. My plants exhibited no tendency to become equal-styled, so as to lose their proper long-styled character, as

73

not rarely happens under cultivation with several heterostyled species of Primula; but it would appear that they had been greatly affected in function, either by long-continued cultivation or by some other cause. We shall see in a future chapter that heterostyled plants illegitimately / fertilized during several successive generations sometimes become more self-fertile; and this may have been the case with my stock of the present species of Pulmonaria; but in this case we must assume that the long-styled plants were at first sufficiently fertile to yield some seed, instead of being ansolutely self-sterile like the German plants.

Pulmonaria angustifolia. Seelings of this plant, raised from plants growing wild in the Isle of Wight, were named for me by Dr Hooker. It is so closely allied to the last species, differing chiefly in the shape and spotting of the leaves, that the two have been considered by several eminent botanists – for instance, Bentham – as mere varieties. But, as we shall presently see, good evidence can be assigned for ranking them as distinct. Owing to the doubts on this head, I tried whether the two would mutually fertilize one another. Twelve short-styled flowers of *P. angustifolia* were legitimately fertilized with pollen from long-styled plants of *P. officinalis* (which, as we have just seen, are moderately self-fertile), but they did not produce a single fruit. Thirty-six long-styled flowers of *P. angustifolia* were also illegitimately fertilized during two seaons with pollen from the long-styled *P. officinalis*, but all these flowers dropped off unimpregnated. Had the plants been mere varieties of the same species these illegitimate crosses would probably have yielded some seeds, judging from my success in illegitimately fertilizing the long-styled flowers of *P. officinalis*; and the twelve legitimate crosses, instead of yielding no fruit, would almost certainly have yielded a considerable number, namely, about nine, judging from the results given in the following table (Table 20). Therefore *P. officinalis* and *angustifolia* appear to be good and distinct species, in / conformity with other important functional differences between them, immediately to be described.

The long-styled and short-styled flowers of *P. angustifolia* differ from one another in structure in nearly the same manner as those of *P. officinalis*. But in the accompanying figure a slight bulging of the corolla in the long-styled form, where the anthers are seated, has been overlooked. My son William, who examined a large number of wild plants in the Isle of Wight, observed that the corolla, though variable in size, was generally larger in the long-styled flowers than in the short-

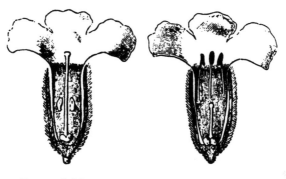

Long-styled form   Short-styled form

Fig. 6   *Pulmonaria angustifolia*

styled; and certainly the largest corollas of all were found on the long-styled plants, and the smallest on the short-styled. Exactly the reverse occurs, according to Hildebrand, with *P. officinalis*. Both the pistils and stamens of *P. angustifolia* vary much in length; so that in the short-styled form the distance between the stigma and the anthers varied from 119 to 65 divisions of the micrometer, and in the long-styled form 115 to 112. From an average of seven / measurements of each form the distance between these organs in the long-styled is to the same distance in the short-styled form as 100 to 69; so that the stigma in the one form does not stand on a level with the anthers in the other. The long-styled pistil is sometimes thrice as long as that of the short-styled; but from an average of ten measurements of both, its length to that of the short-styled was as 100 to 56. The stigma varies in being more or less, though slightly, lobed. The anthers also vary much in length in both forms, but in a greater degree in the long-styled than in the short-styled form; many in the former being from 80 to 63, and in the latter from 80 to 70 divisions of the micrometer in length. From an average of seven measurements, the short-styled anthers were to those from the long-styled as 100 to 91 in length. Lastly, the pollen grains from the long-styled flowers varied between 13 and 11·5 divisions of the micrometer, and those from the short-styled between 15 and 13. The average diameter of 25 grains from the latter, or short-styled form, was to that of 20 grains from the long-styled as 100 to 91. We see, therefore, that the pollen grains from the smaller anthers of the shorter stamens in the long-styled form are, as usual, of smaller size

75

than those in the other form. But what is remarkable, a larger proportion of the grains were small, shrivelled, and worthless. This could be seen by merely comparing the contents of the anthers from several distinct plants of each form. But in one instance my son found, by counting, that out of 193 grains from a long-styled flower, 53 were bad, or 27 per cent; whilst out of 265 grains from a short-styled flower only 18 were bad, or 7 per cent. From the condition of the pollen in the long-styled form, and from the extreme variability / of all the organs in both forms, we may perhaps suspect that the plant is undergoing a change, and tending to become dioecious.

My son collected in the Isle of Wight on two occasions 202 plants, of which 125 were long-styled and 77 short-styled; so that the former were the more numerous. On the other hand, out of 18 plants raised by me from seed, only 4 were long-styled and 14 short-styled. The short-styled plants seemed to my son to produce a greater number of flowers than the long-styled; and he came to this conclusion before a similar statement had been published by Hildebrand with respect to *P. officinalis*. My son gathered ten branches from ten different plants of both forms, and found the number of flowers of the two forms to be as 100 to 89, 190 being short-styled and 169 long-styled. With *P. officinalis* the difference, according to Hildebrand, is even greater, namely, as 100 flowers for the short-styled to 77 for the long-styled plants. The following table shows the results of my experiments:

TABLE 20   *Pulmonaria angustifolia*

Nature of union	Number of flowers fertilized	Number of fruits produced	Average number of seeds per fruit
Long-styled flowers, by pollen of short-styled. Legitimate union	18	9	2·11
Long-styled flowers, by own-form pollen. Illegitimate union	18	0	0
Short-styled flowers, by pollen of long-styled. Legitimate union	18	15	2·60
Short-styled flowers, by own-form pollen. Illegitimate union	12	7	1·86 /

We see in this table that the fertility of the two legitimate unions to that of the two illegitimate together is as 100 to 35, judged by the proportion of flowers which produced fruit: and as 100 to 32, judged by the average number of seeds per fruit. But the small number of fruit yielded by the 18 long-styled flowers in the first line was probably accidental, and if so, the difference in the proportion of legitimately and illegitimately fertilized flowers which yield fruit is really greater than that represented by the ratio of 100 to 35. The 18 long-styled flowers illegitimately fertilised yielded no seeds – not even a vestige of one. Two long-styled plants which were placed under a net produced 138 flowers, besides those which were artificially fertilized, and none of these set any fruit; nor did some plants of the same form which were protected during the next summer. Two other long-styled plants were left uncovered (all the short-styled plants having been previously covered up), and humble-bees, which had their foreheads white with pollen, incessantly visited the flowers, so that their stigmas must have received an abundance of pollen, yet these flowers did not produce a single fruit. We may therefore conclude that the long-styled plants are absolutely barren with their own-form pollen, though brought from a distinct plant. In this respect they differ greatly from the long-styled English plants of *P. officinalis* which were found by me to be moderately self-fertile; but they agree in their behaviour with the German plants of *P. officinalis* experimented on by Hildebrand.

Eighteen short-styled flowers legitimately fertilized yielded, as may be seen in Table 20, 15 fruits, each having on an average 2·6 seeds. Four of these fruits contained the highest possible number of seeds, namely / 4, and four other fruits contained each 3 seeds. The 12 illegitimately fertilized short-styled flowers yielded 7 fruits, including on an average 1·86 seeds; and one of these fruits contained the maximum number of 4 seeds. This result is very surprising in contrast with the absolute barrenness of the long-styled flowers when illegitimately fertilized; and I was thus led to attend carefully to the degree of self-fertility of the short-styled plants. A plant belonging to this form and covered by the net bore 28 flowers besides those which had been artificially fertilized, and of all these only two produced a fruit each including a single seed. This high degree of self-sterility no doubt depended merely on the stigmas not receiving any pollen, or not a sufficient quantity. For after carefully covering all the long-styled plants in my garden, several short-styled plants were left exposed to the visits of humble-bees, and their stigmas will thus have received plenty of short-styled pollen; and now about half the flowers, thus illegitimately fertilized, set fruit. I judge of this proportion partly from estimation and party from having examined three large branches, which had borne 31 flowers, and these produced 16 fruits. Of the fruits produced 233 were collected (many being left ungathered), and these included on an average 1·82 seeds. No less than 16 out of the 233 fruits included the highest possible number of seeds, namely 4, and 31 included 3 seeds. So we see how highly fertile these short-styled plants were when illegitimately fertilized with their own-form pollen by the aid of bees.

The great difference in the fertility of the long- and short-styled flowers, when both are illegitimately fertilized, is a unique case, as far as I have observed with heterostyled plants. The long-styled flowers when thus fertilized are utterly barren, whilst about half of the / short-styled ones produce capsules, and these include a little above two-thirds of the number of seeds yielded by them when legitimately fertilized. The sterility of the illegitimately fertilized long-styled flowers is probably increased by the deteriorated condition of their pollen; nevertheless this pollen was highly efficient when applied to the stigmas of the short-styled flowers. With several species of Primula the short-styled flowers are much more sterile than the long-styled, when both are illegitimately fertilized; and it is a tempting view, as formerly remarked, that this greater sterility of the short-styled flowers is a special adaptation to check self-fertilization, as their stigmas are eminently liable to receive their own pollen. This view is even still more tempting in the case of the long-styled form of *Linum grandiflorum*. On the other hand, with *Pulmonaria augustifolia*, it is evident, from the corolla projecting obliquely upwards, that pollen is much more likely to fall on, or to be carried by insects down to the stigma of the short-styled than of the long-styled flowers; yet the short-styled instead of being more sterile, as a protection against self-fertilization, are far more fertile than the long-styled, when both are illegitimately fertilized.

Pulmonaria azurea, according to Hildebrand, is not heterostyled.[12]

From an examination of dried flowers of *Amsinckia spectabilis*, sent to me by Professor Asa Gray, I formerly thought that this plant, a member of the Boragineae, was heterostyled. The pistil varies to an extraordinary degree in length, being in some specimens twice as long as in others, and the point of insertion of the stamens likewise varies. But on raising many plants from seed, I soon became convinced that the whole case was one of mere variability. The first-formed flowers are apt to / have stamens somewhat arrested in develop-

---

[12] *Die Geschlechter-Vertheilung bei den Pflanzen*, 1867, p. 37.

ment with very little pollen in their anthers; and in such flowers the stigma projects above the anthers, whilst generally it stands below and sometimes on a level with them. I could detect no difference in the size of the pollen grain or in the structure of the stigma in the plants which differed most in the above respects; and all of them, when protected from the access of insects, yielded plenty of seeds. Again, from statements made by Vaucher, and from a hasty inspection, I thought at first that the allied *Anchusa arvensis* and *Echium vulgare* were heterostyled, but soon saw my error. From information given me, I examined dried flowers of another member of the Boragineae, *Arnebia hispidissima*, collected from several sites, and though the corolla, together with the included organs, differed much in length, there was no sign of heterostylism.

POLYGONUM FAGOPYRUM (POLYGONACEAE)

Hildebrand has shown that this plant, the common buck-wheat, is heterostyled.[13] In the long-styled form (Fig. 7), the three stigmas project considerably above the eight short stamens, and stand on a level

Upper figure, the long-styled form; lower figure, the short-styled. Some of the anthers have dehisced, others have not.

Fig. 7   *Polygonum fagopyrum (from H. Müller)*

[13] *Die Geschlechter-Vertheilung*, etc., 1867, p. 34.

79

with the anthers of the eight long stamens in the short-styled form; and so it is conversely with the stigmas and stamens of this latter form. I could perceive no difference in the structure of the stigmas in the two forms. The pollen grains of the short-styled form are to those of the long-styled as 100 to 82 in diameter. This plant is therefore without doubt heterostyled.

I experimented only in an imperfect manner on the relative fertility of the two forms. Short-styled flowers were dragged several times over two heads of flowers on long-styled plants, protected under a net, which was thus legitimately, though not fully, fertilized. They produced 22 seeds, or 11 per flower-head.

Three flower-heads on long-styled plants received / pollen in the same manner from other long-styled plants, and were thus illegitimately fertilized. They produced 14 seeds, or only 4·66 per flower-head.

Two flower-heads on short-styled plants received pollen in like manner from long-styled flowers, and were thus legitimately fertilized. They produced 8 seeds, or 4 per flower-head.

Four heads on short-styled plants similarly received pollen from other short-styled plants, and were thus illegitimately fertilized. They produced 9 seeds, or 2·25 per flower-head.

The result from fertilizing the flower-heads in the above imperfect manner cannot be fully trusted; but I may state that the four legitimately fertilized flower-heads / yielded on an average 7·50 seeds per head; whereas the seven illegitimately fertilized heads yielded less than half the number, or on an average only 3·28 seeds. The legitimately crossed seeds from the long-styled flowers were finer than those from the illegitimately fertilized flowers on the same plant, in the ratio of 100 to 82, as shown by the weights of an equal number.

About a dozen plants, including both forms, were protected under nets, and early in the season they produced spontaneously hardly any seeds, though at this period the artificially fertilized flowers produced an abundance; but it is a remarkable fact that later in the season, during September, both forms became highly self-fertile. They did not, however, produce so many seeds as some neighbouring uncovered plants which were visited by insects. Therefore the flowers of neither form when left to fertilize themselves late in the season without the aid of insects, are nearly so sterile as most other heterostyled plants. A large number of insects, namely 41 kinds as observed by H. Müller,[14] visit the

[14] *Die Befruchtung*, etc., p. 175, and *Nature*, 1 January, 1874, p. 166.

flowers for the sake of the eight drops of nectar. He infers from the structure of the flowers that insects would be apt to fertilize them both illegitimately as well as legitimately; but he is mistaken in supposing that the long-styled flowers cannot spontaneously fertilize themselves.

Differently to what occurs in the other genera hitherto noticed, Polygonum, though a very large genus, contains, as far as is at present known, only a single heterostyled species, namely the present one. H. Müller in his interesting description of several / other species shows that *P. bistorta* is so strongly proterandrous (the anthers generally falling off before the stigmas are mature) that the flowers must be cross-fertilized by the many insects which visit them. Other species bear much less conspicuous flowers which secrete little or no nectar, and consequently are rarely visited by insects; these are adapted for self-fertilization, though still capable of cross-fertilization. According to Delpino, the Polygonaceae are generally fertilized by the wind, instead of by insects as in the present genus.

LEUCOSMIA BURNETTIANA (THYMELIAE)

As Professor Asa Gray has expressed his belief[15] that this species and *L. acuminata*, as well as some species in the allied genus Drymispermum, are dimporphic or heterostyled, I procured from Kew, through the kindness of Dr Hooker, two dried flowers of the former species, an inhabitant of the Friendly Islands in the Pacific. The pistil of the long-styled form is to that of the short-styled as 100 to 86 in length; the stigma projects just above the throat of the corolla, and is surrounded by five anthers, the tips of which reach up almost to its base; and lower down, within the tubular corolla, five other and rather smaller anthers are seated. In the short-styled form, the stigma stands some way down the tube of the corolla, nearly on a level with the lower anthers of the other form: it differs remarkably from the stigma of the long-styled form, it being more papillose, and in being longer in the ratio of 100 to 60. The anthers of the upper stamens in the short-styled form are supported on free filaments, and project above the throat of the corolla, whilst the anthers of the lower stamens are seated in the throat on a level with the upper stamens of the other form. The diameters of a considerable number of grains from both sets of anthers in both forms were measured, but they did not differ in any trusthworthy degree. The mean diameter of twenty-two / grains from the short-styled flower was to that of twenty-four grains from the long-styled, as 100 to 99. The anthers of the upper stamens in the short-styled form appeared to be poorly developed, and contained a considerable number of shrivelled grains which were omitted in striking the above average. Notwithstanding the

[15] *American Journal of Science*, 1865, p. 101, and Seemann's *Journal of Botany*, vol. iii, 1865, p. 305.

fact of the pollen-grains from the two forms not differing in diameter in any appreciable degree, there can hardly be a doubt from the great difference in the two forms in the length of the pistil, and especially of the stigma, together with its more papillose condition in the short-styled form, that the present species is truly heterostyled. This case resembles that of *Linum grandiflorum*, in which the sole difference between the two forms consists in the length of the pistils and stigmas. From the great length of the tubular corolla of Leucosmia, it is clear that the flowers are cross-fertilized by large Lepidoptera or by honey-sucking birds, and the position of the stamens in two whorls one beneath the other, which is a character that I have not seen in any other heterostyled dimorphic plant, probably serves to smear the inserted organ thoroughly with pollen.

MENYANTHES TRIFOLIATA (GENTIANEAE)

This plant inhabits marshes: my son William gathered 247 flowers from so many distinct plants, and of these 110 were long-styled, and 137 short-styled. The pistil of the long-styled form is in length to that of the short-styled in the ratio of about 3 to 2. The stigma of the former, as my son observed, is decidedly larger than that of the short-styled; but in both forms it varies much in size. The stamens of the short-styled are almost double the length of those of the long-styled; so that their anthers stand rather above the level of the stigma of the long-styled form. The anthers also vary much in size, but seem often to be of larger size in the short-styled flowers. My son made with the camera many drawings of the pollen-grains, and those from the short-styled flowers were in diameter in nearly the ratio of 100 to 84 to those from the long-styled flowers. I know nothing about the capacity for fertilization in the two forms; but short-styled plants, living by themselves in the gardens at Kew, have produced an abundance of capsules, yet the seeds have never germinated; and this looks as if the short-styled form was sterile with its own pollen. /

LIMNANTHEMUM INDICUM (GENTIANEAE)

This plant is mentioned by Mr Thwaites in his *Enumeration of the Plants of Ceylon* as presenting two forms; and he was so kind as to send me specimens preserved in spirits. The pistil of the long-styled form is nearly thrice as long (i.e. as 14 to 5) as that of the short-styled, and is very much thinner in the ratio of about 3 to 5. The foliaceous stigma is more expanded, and twice as large as that of the short-styled form. In the latter the stamens are about twice as long as those of the long-styled, and their anthers are longer in the ratio of 100 to 70. The pollen grains, after having been long kept in spirits, were of the same shape and size in both forms. The ovules, according to Mr Thwaites, are equally numerous (viz. from 70 to 80) in the two forms.

82

VILLARSIA [SP.?] (GENTIANEAE)

Fritz Müller sent me from South Brazil dried flowers of this aquatic plant, which is closely allied to Limnanthemum. In the long-styled form the stigma stands some way above the anthers, and the whole pistil, together with the ovary, is in length to that of the short-styled form as about 3 to 2. In the latter form the anthers stand above the stigma, and the style is very short and thick; but the pistil varies a good deal in length, the stigma being either on a level with the tips of the sepals or considerably beneath them. The foliaceous stigma in the long-styled form is larger, with the expansions running farther down the style, than in the other form. One of the most remarkable differences between the two forms is that the anthers of the longer stamens in the short-styled flowers are conspicuously longer than those of the shorter stamens in the long-styled flowers. In the former the subtriangular pollen grains are larger; the ratio between their breadth (measured from one angle to the middle of the opposite side) and that of the grains from the long-styled flowers being about 100 to 75. Fritz Müller also informs me that the pollen of the short-styled flowers has a bluish tint, whilst that of the long-styled is yellow. When we treat of *Lythrum salicaria* we shall find a strongly marked contrast in the colour of the pollen in two of the forms.

The three genera, Menyanthes, Limnanthemum, and Villarsia, now described, constitute a well-marked subtribe of the Gentianeae. All the species, as far as at present known, are heterostyled, and all inhabit aquatic or sub aquatic stations. /

FORSYTHIA SUSPENSA (OLEACEAE)

Professor Asa Gray states that the plants of this species growing in the Botanic Gardens at Cambridge, U.S., are short-styled, but that Siebold and Zuccarini describe the long-styled form, and give figures of two forms; so that there can be little doubt, as he remarks, about the plant being dimorphic.[16] I therefore applied to Dr Hooker, who sent me a dried flower from Japan, another from China, and another from the Botanic Gardens at Kew. The first proved to be long-styled, and the other two short-styled. In the long-styled form, the pistil is in length to that of the short-styled as 100 to 38, the lobes of the stigma being a little longer (as 10 to 9), but narrower and less divergent. This last character, however, may be only a temporary one. There seems to be no difference in the papillose condition of the two stigmas. In the short-styled form, the stamens are in length to those of the long-styled as 100 to 66, but the anthers are shorter in the ratio of 87 to 100; and this is unusual, for when there is any difference in size between the anthers of the two forms, those from the longer stamens of the short-styled are generally the longest. The pollen grains from the short-styled flowers are certainly larger, but only in a slight degree, than those from the

[16] *The American Naturalist*, July, 1873, p. 422.

long-styled, namely, as 100 to 94 in diameter. The short-styled form which grows in the Gardens at Kew, has never there produced fruit.

Forsythia viridissima appears likewise to be heterostyled; for Professor Asa Gray says that although the long-styled form alone grows in the gardens at Cambridge, U.S., the published figures of this species belong to the short-styled form.

CORDIA [SP.?] (CORDIACEAE)

Fritz Müller sent me dried specimens of this shrub, which he believes to be heterostyled; and I have not much doubt that this is the case, though the usual characteristic differences are not well pronounced in the two forms. *Linum grandiflorum* shows us that a plant may be heterostyled in function in the highest degree, and yet the two forms may have stamens of equal length, and pollen grains of equal size. In the present species of Cordia, the stamens of both forms are of nearly equal / length, those of the short-styled being rather the longest; and the anthers of both are seated in the mouth of the corolla. Nor could I detect any difference in the size of the pollen-grains, when dry or after being soaked in water. The stigmas of the long-styled form stand clear above the anthers, and the whole pistil is longer than that of the short-styled, in about the ratio of 3 to 2.

The stigmas of the short-styled form are seated beneath the anthers, and they are considerably shorter than those of the long-styled form. This latter difference is the most important one of any between the two forms.

GILIA (IPOMOPSIS) *PULCHELLA VEL AGGREGATA* (POLEMONIACEAE)

Professor Asa Gray remarks with respect to this plant: 'the tendency to dimorphism, of which there are traces, or perhaps rather incipient manifestations in various portions of the genus, is most marked in *G. aggregata*'.[17] He sent me some dried flowers, and I procured others from Kew. They differ greatly in size, some being nearly twice as long as others (viz. as 30 to 17), so that it was not possible to compare, except by calculation, the absolute length of the organs from different plants. Moreover, the relative position of the stigmas and anthers is variable: in some long-styled flowers the stigmas and anthers were exserted only just beyond the throat of the corolla: whilst in others they were exserted as much as 4/10 of an inch. I suspect also that the pistil goes on growing for some time after the anthers have dehisced. Nevertheless it is possible to class the flowers under two forms. In some of the long-styled, the length of pistil to that of the short-styled was as 100 to 82; but this result was gained by reducing the size of the corollas to the same scale. In another pair of flowers the difference in length between the pistils of the two forms was certainly greater, but they were not actually measured. In the short-styled flowers, whether large or small, the stigma is seated low down within the tube

---

[17] *Proc. American Acad. of Arts and Sciences*, 14 June, 1870, p. 275.

of the corolla. The papillae on the long-styled stigma are longer than those on the short-styled, in the ratio of 100 to 40. The filaments in some of the short-styled flowers were, to those of the long-styled, as 100 to 25 in length, the free, or / unattached portion being alone measured; but this ratio cannot be trusted, owing to the great variability of the stamens. The mean diameter of eleven pollen grains from long-styled flowers, and of twelve from the short-styled, was exactly the same. It follows from these several statements, that the difference in length and state of surface of the stigmas in the flowers is the sole reliable evidence that this species is heterostyled; for it would be rash to trust to the difference in the length of the pistils, seeing how variable they are. I should have left the case altogeher doubtful, had it not been for the observations on the following species; and these leave little doubt on my mind that the present plant is truly heterostyled. Professor Gray informs me that in another species, *G. coronopifolia*, belonging to the same section of the genus, he can see no sign of dimorphism.

GILIA (LEPTOSIPHON) *MICRANTHA*

A few flowers sent me from Kew had been somewhat injured, so that I cannot say anything positively with respect to the position and relative length of the organs in the two forms. But their stigmas differed almost exactly in the same manner as in the last species; the papillae on the long-styled stigma being longer than those on the short-styled, in the ratio of 100 to 42. My son measured nine pollen grains from the long-styled, and the same number from the short-styled form; and the mean diameter of the former was to that of the latter as 100 to 81. Considering this difference, as well as that between the stigmas of the two forms, there can be no doubt that this species is heterostyled. So probably is *Gilia nudicaulis*, which likewise belongs to the Leptosiphon section of the genus, for I hear from Professor Asa Gray that in some individuals the style is very long, with the stigma more or less exserted, whilst in others it is deeply included within the tube; the anthers being always seated in the throat of the corolla.

PHLOX SUBULATA (POLEMONIACEAE)

Professor Asa Gray informs me that the greater number of the species in this genus have a long pistil, with the stigma more or less exserted; whilst several other species, especially the annuals, have a short pistil seated low down within the tube of the corolla. In all the species the anthers are arranged one / below the other, the uppermost just protruding from the throat of the corolla. In *Phlox subulata* alone he has 'seen both long and short styles; and here the short-styled plant has (irrespective of this character) been described as a distinct species (*P. nivalis*, *P. Hentzii*), and is apt to have a pair of ovules in each cell, while the long-styled *P. subulata* rarely shows more than one'.[18] Some dried

[18] *Proc. American Acad. of Arts and Sciences*, 14 June, 1870, p. 248.

flowers of both forms were sent me by him, and I received others from Kew, but I have failed to make out whether the species is heterostyled. In two flowers of nearly equal size, the pistil of the long-styled form was twice as long as that of the short-styled; but in other cases the difference was not nearly so great. The stigma of the long-styled pistil stands nearly in the throat of the corolla; whilst in the short-styled it is placed low down – sometimes very low down in the tube, for it varies greatly in position. The stigma is more papillose, and of greater length (in one instance in the ratio of 100 to 67), in the short-styled flowers than in the long-styled. My son measured twenty pollen grains from a short-styled flower, and nine from a long-styled, and the former were in diameter to the latter as 100 to 93; and this difference accords with the belief that the plant is heterostyled. But the grains from the short-styled varied much in diameter. He afterwards measured ten grains from a distinct long-styled flower, and ten from another plant of the same form, and these grains differed in diameter in the ratio of 100 to 90. The mean diameter of these two lots of twenty grains was to that of twelve grains from another short-styled flower as 100 to 75: here, then, the grains from the short-styled form were considerably smaller than those from the long-styled, which is the reverse of what occurred in the former instance, and of what is the general rule with heterostyled plants. The whole case is perplexing in the highest degree, and will not be understood until experiments are tried on living plants. The greater length and more papillose condition of the stigma in the short-styled than in the long-styled flowers, looks as if the plant was heterostyled; for we know that with some species – for instance, Leucosmia and certain Rubiaceae – the stigma is longer and more papillose in the short-styled form, though the reverse of this holds good in Gilia, a member of the same family with Phlox. The similar position of the anthers in the two forms is somewhat / opposed to the present species being heterostyled; as is the great difference in the length of the pistil in several short-styled flowers. But the extraordinary variability in diameter of the pollen grains, and the fact that in one set of flowers the grains from the long-styled flowers were larger than those from the short-styled, is strongly opposed to the belief that *Phlox subulata* is heterostyled. Possibly this species was once heterostyled, but is now becoming sub-dioecious; the short-styled plants having been rendered more femine in nature. This would account for their ovaries usually containing more ovules, and for the variable condition of their pollen grains. Whether the long-styled plants are now changing their nature, as would appear to be the case from the variability of their pollen grains, and are becoming more masculine, I will not pretend to conjecture; they might remain as hermaphrodites, for the co-existence of hermaphrodite and female plants of the same species is by no means a rare event.

ERYTHROXYLUM [SP. ?] (ERYTHROXYLIDAE)

Fritz Müller sent me from South Brazil dried flowers of this tree, together with the accompanying drawings, which show the two forms, magnified about five times, with the petals removed. / In the long-styled form the stigmas project above the anthers, and the styles are nearly twice as long as those of the short-

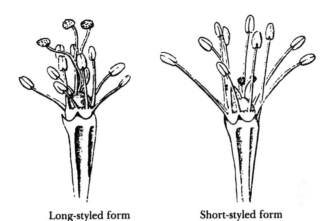

Long-styled form   Short-styled form

From a sketch by Fritz Müller, magnified five times

Fig. 8   Erythroxylon [sp. ?]

styled form, in which the stigmas stand beneath the anthers. The stigmas in many, but not in all the short-styled flowers, are larger than those in the long-styled. The anthers of the short-styled flowers stand on a level with the stigmas of the other form; but the stamens are longer by only one-fourth or one-fifth of their own length than those of the long-styled. Consequently the anthers of the latter do not stand on a level with, but rather above the stigmas of the other form. Differently from what occurs in the following closely allied genus, Sethia, the stamens are of nearly equal length in the flowers of the same form. The pollen grains of the short-styled flowers, measured in their dry state, are a little larger than those from the long-styled flowers in about the ratio of 100 to 93.[19]

SETHIA ACUMINATA (ERYTHROXYLIDAE)

Mr Thwaites pointed out several years ago[20] that this plant exists under two forms, which he designated as *forma stylosa et staminea*; and the flowers sent to me by him are clearly heterostyled. In the long-styled form the pistil is nearly twice as long, and the stamens half as long as the corresponding organs in the short-styled form. The stigmas of the long-styled seem rather smaller than those of the short-styled. All the stamens in the short-styled flowers are of

[19] F. Müller remarks in his letter to me that the flowers, of which he carefully examined many specimens, are curiously variable in the number of their parts: 5 sepals and petals, 10 stamens and 3 pistils are the prevailing numbers; but the sepals and petals often vary from 5 to 7; the stamens from 10 to 14, and the pistils from 3 to 4.
[20] *Enumeratio Plantarum Zeylaniae*, 1864, p. 54.

nearly equal length, whereas in long-styled they differ in length, being alternately a little longer and shorter; and this difference in the stamens of the two forms is probably related, as we shall hereafter see in the case of the short-styled flowers of *Lythrum salicaria*, to the manner in which insects can best transport pollen from the long-styled flowers to the stigmas of the short-styled. The pollen grains from the short-styled flowers, though variable in size, are to those of the long-styled, as far as I could make out, as 100 to 83 in their longer diameter. *Sethia obtusifolia* is heterostyled like *S. acuminata*. /

CRATOXYLON FORMOSUM (HYPERICINEAE)

Mr Thiselton Dyer remarks that this tree, an inhabitant of Malacca and Borneo, appears to be heterostyled.[21] He sent me dried flowers, and the difference between the two forms is conspicuous. In the long-styled form the pistils are in length to those of the short-styled as 100 to 40, with their globular stigmas about twice as thick. These stand just above the numerous anthers and a little beneath the tips of the petals. In the short-styled form the anthers project high above the pistils, the stigmas of which diverge between the three bundles of stamens, and stand only a little above the tips of the sepals. The stamens in this form are to those of the long-styled as 100 to 86 in length; and therefore they do not differ so much in length as do the pistils. Ten pollen grains from each form were measured and those from the short-styled were to those from the long-styled as 100 to 86 in diameter. This plant, therefore, is in all respects a well-characterized heterostyled species.

AEGIPHILA ELATA (VERBENACEAE)

Mr Bentham was so kind as to send me dried flowers of this species and of *Ae. mollis*, both inhabitants of South America. The two forms differ conspicuously, as the deeply bifid stigma of the one, and the anthers of the other project far above the mouth of the corolla. In the long-styled form of the present species, the style is twice and a half as long as that of the short-styled. The divergent stigmas of the two forms do not differ much in length, nor as far as I could perceive in their papillae. In the long-styled flowers the filaments adhere to the corolla close up to the anthers, which are enclosed some way down within the tube. In the short-styled flowers the filaments are free above the point where the anthers are seated in the other form, and they project from the corolla to an equal height with that of the stigmas in the long-styled flowers. It is often difficult to measure with accuracy pollen grains, which have long been dried and then soaked in water; but they here manifestly differed greatly in size. Those from the short-styled flowers were to those from the long-styled in diameter in / about the ratio of 100 to 62. The two forms of *Ae. mollis* present a like difference in the length of their pistils and stamens.

[21] *Journal of Botany*, London, 1872, p. 26.

AEGIPHILA OBDURATA

Flowers of this bush were sent me from St Catharina in Brazil, by Fritz Müller, and were named for me at Kew. They appeared at first sight grandly hetero-styled, as the stigma of the long-styled form projects far out of the corolla, whilst the anthers are seated halfway down within the tube; whereas in the short-styled form the anthers project from the corolla and the stigma is enclosed in the tube at nearly the same level with the anthers of the other form. The pistil of the long-styled is to that of the short-styled as 100 to 60 in length, and the stigmas, taken by themselves, as 100 to 55. Nevertheless, this plant cannot be heterostyled. The anthers in the long-styled form are brown, tough, and fleshy, and less than half the length of those in the short-styled form, strictly as 44 to 100; and what is much more important, they were in a rudimentary condition in the two flowers examined by me, and did not contain a single grain of pollen. In the short-styled form, the divided stigma, which as we have seen is much shortened, is thicker and more fleshy than the stigma of the long-styled, and is covered with small irregular projections, formed of rather large cells. It had the appearance of having suffered from hypertrophy, and is probably in-capable of fertilization. If this be so the plant is dioecious, and judging from the two species previously described, it probably was once heterostyled, and has since been rendered dioecious by the pistil in the one form, and the stamens in the other having become functionless and reduced in size. It is, however, possible that the flowers may be in the same state as those of the common thyme and of several other Labiatae, in which females and hermaphrodites regularly co-exist. Fritz Müller, who thought that the present plant was heterostyled, as I did at first, informs me that he found bushes in several places growing quite isolated, and that these were completely sterile; whilst two plants growing close together were covered with fruit. This fact agrees better with the belief that the species is dioecious than that it consists of hermaphrodites and females; for if anyone of the isolated plants had been an hermaphrodite, it would probably have produced some fruit. /

RUBIACEAE

This great natural family contains a much larger number of hetero-styled genera than any other one, as yet known.

Mitchella repens. Professor Asa Gray sent me several living plants collected when out of flower, and nearly half of these proved long-styled, and the other half short-styled. The white flowers, which are fragrant and which secrete plenty of nectar, always grow in pairs with their ovaries united, so that the two together produce 'a berry-like double drupe'.[22] In my first series of experiments (1864) I did not suppose that this curious arrangement of the flowers would have any influence on their fertility; and in several instances only one of the two flowers in a pair was fertilized; and a large proportioin or all of these failed to produce berries. In the ensuing year both flowers of each pair were invariably fertilized in the same manner; and the latter experiments alone serve to show the proportion of flowers which yield berries, when legitimately and illegitimately fertilized; but for calculating the average number of seeds per berry I have used those produced during both seasons.

In the long-styled flowers the stigma projects just above the bearded throat of the corolla, and the anthers are seated some way down the tube. In the short-styled flowers these organs occupy reversed positions. In this latter form the fresh pollen grains are a little larger and more opaque than those of the long-styled form. The results of my experiments are given in the following table. /

TABLE 21   *Mitchella repens*

Nature of union	Number of pairs of flowers fertilised during the second season	Number of drupes produced during the second season	Average number of good seeds per drupe in all the drupes during the two seasons
Long-styled flowers, by pollen of short-styled. Legitimate union	9	8	4·6
Long-styled flowers, by own-form pollen. Illegitimate union	8	3	2·2
Short-styled flowers, by pollen of long-styled. Legitimate union	8	7	4·1
Short-styled flowers, by own-form pollen. Illegitimate union	9	0	2·0
The two legitimate unions together	17	15	4·4
The two illegitimate unions together	17	3	2·1

It follows from this table that 88 per cent of the paired flowers of both forms, when legitimately fertilized, yielded double berries, nineteen of which contained on an average 4·4 seeds, with a maximum in one of 8 seeds. Of the illegitimately fertilized paired flowers only 18

[22] A Gray, *Manual of the Bot. of the N. United States*, 1856, p. 172.

per cent yielded berries, six of which contained on an average only 2·1 seeds, with a maximum in one of 4 seeds. Thus the two legitimate unions are more fertile than the two illegitimate, according to the proportion of flowers which yielded berries, in the ratio of 100 to 20; and according to the average number of contained seeds as 100 to 47.

Three long-styled and three short-styled plants were protected under separate nets, and they produced altogether only 8 berries, containing on an average only / 1·5 seeds. Some additional berries were produced which contained no seeds. The plants thus treated were therefore excessively sterile, and their slight degree of fertility may be attributed in part to the action of the many indivuals of thrips which haunted the flowers. Mr J. Scott informs me that a single plant (probably a long-styled one), growing in the Botanic Gardens at Edinburgh, which no doubt was freely visited by insects, produced plenty of berries, but how many of them contained seeds was not observed.

BORRERIA, NOV. SP. NEAR VALEROANOIDES (RUBIACEAE)

Fritz Müller sent me seeds of this plant, which is extremely abundant in St Catharina, in South Brazil; and ten plants were raised, consisting of five long-styled and five short-styled. The pistil of the long-styled flowers projects just beyond the mouth of the corolla, and is thrice as long as that of the short-styled, and the divergent stigmas are likewise rather larger. The anthers in the long-styled form stand low down within the corolla, and are quite hidden. In the short-styled flowers the anthers project just above the mouth of the corolla, and the stigma stands low down within the tube. Considering the great difference in the length of the pistils in the two forms, it is remarkable that the pollen grains differ very little in size, and Fritz Müller was struck with the same fact. In a dry state the grains from the short-styled flowers could just be perceived to be larger than those from the long-styled, and when both were swollen by immersion in water, the former were to the latter in diameter in the ratio of 100 to 92. In the long-styled flowers beaded hairs almost fill up the mouth of the corolla and project above it; they therefore stand above the anthers and beneath the stigma. In the / short-styled flowers a similar brush of hairs is situated low down within the tubular corolla, above the stigma and beneath the anthers. The presence of these beaded hairs in both forms, though occupying such different positions, shows that they are probably of considerable functional importance. They would serve to guard the stigma of each form from its own pollen; but in accordance with Professor Kerner's view[23] their chief use probably is to prevent the copious nectar being stolen by small crawling insects, which could not render any service to the species by carrying pollen from one form to the other.

The flowers are so small and so crowded together that I was not willing to expend time in fertilizing them separately; but I dragged repeatedly heads of short-styled flowers over three long-styled flower-heads, which were thus legitimately fertilized; and they produced many dozen fruits, each containing two good seeds. I fertilized in the same manner three heads on the same long-styled plant with pollen from another long-styled plant, so that these were fertilized illegitimately, and they did not yield a single seed. Nor did this plant, which was of course protected by a net, bear spontaneously any seeds. Nevertheless another long-styled plant, which was carefully protected, produced spontaneously a very few seeds; so that the long-styled form is not always quite sterile with its own pollen.

FARAMEA [SP.?] (RUBIACEAE)

Fritz Müller has fully described the two forms of this remarkable plant, an inhabitant of South Brazil.[24] In / the long-styled form the pistil projects above the corolla, and is almost exactly twice as long as that of the short-styled, which is included within the tube. The former is divided into two rather short and broad stigmas, whilst the short-styled pistil is divided into two long, thin, sometimes much curled stigmas. The stamens of each form correspond in height or length with the pistils of the other form. The anthers of the short-styled form are a little larger than those of the long-styled; and their pollen grains are to those of the other form as 100 to 67 in diameter. But the pollen grains of the two forms differ in a much more remarkable manner, of which no other instance is known; those from the short-styled flowers being covered with sharp points; the smaller ones / from the long-styled being quite smooth. Fritz Müller remarks that this difference between

[23] *Die Schutzmittel der Blüthen gegen unberufene Gäste*, 1876, p. 37.
[24] *Bot. Zeitung*, 10 September, 1869, p. 606.

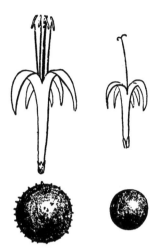

Short-styled form   Long-styled form

Outlines of flowers from dried specimens. Pollen-grains, magnified 180 times, by
Fritz Müller

Fig. 9   Faramea [sp.?]

the pollen grains of the two forms is evidently of service to the plant;
for the grains from the projecting stamens of the short-styled form if
smooth, would have been liable to be blown away by the wind, and
would thus have been lost; but the little points on their surfaces cause
them to cohere, and at the same time favour their adhesion to the
hairy bodies of insects, which merely brush against the anthers of these
stamens whilst visiting the flowers. On the other hand, the smooth
grains of the long-styled flowers are safely included within the tube of
the corolla, so that they cannot be blown away, but are almost sure to
adhere to the proboscis of an entering insect, which is necessarily
pressed close against the enclosed anthers.

It may be remembered that in the long-styled form of *Linum perenne*
each separate stigma rotates on its own axis, when the flower is mature,
so as to turn its papillose surface outwards. There can be no doubt that
this movement, which is confined to the long-styled form, is effected in
order that the proper surface of the stigma should receive pollen
brought by insects from the other form. Now with Faramea, as Fritz
Müller shows, it is the stamens which rotate on their axes in one of the
two forms, namely, the short-styled, in order that their pollen should

be brushed off by insects and transported to the stigmas of the other form. In the long-styled flowers the anthers of the short enclosed stamens, do not rotate on their axes, but dehisce on their inner sides, as is the common rule with the Rubiaceae; and this is the best position for the adherence of the pollen grains to the proboscis of an entering insect. Fritz Müller therefore infers that as the plant became hetero-styled, and as the / stamens of the short-styled form increased in length, they gradually acquired the highly beneficial power of rotating on their own axes. But he has further shown, by the careful examination of many flowers, that this power has not as yet been perfected; and, consequently, that a certain proportion of the pollen is rendered useless, namely, that from the anthers which do not rotate properly. It thus appears that the development of the plant has not as yet been completed; the stamens have indeed acquired their proper length, but not their full and perfect power of rotation.[25]

The several points of difference in structure between the two forms of Faramea are highly remarkable. Until within the recent period, if anyone had been shown two plants which differed in a uniform manner in the length of their stamens and pistils – in the form of their stigmas – in the manner of dehiscence and slightly in the size of their anthers – and to an extraordinary degree in the diameter and structure of their pollen grains, he would have declared it impossible that the two could have belonged to one and the same species.

SUTERIA (SPECIES UNNAMED IN THE HERBARIUM AT KEW) (RUBIACEAE)

I owe to the kindness of Fritz Müller dried flowers of this plant from St Catharina, in Brazil. In the long-styled form the stigma stands in the mouth of the corolla, above the anthers / which latter are enclosed within the tube, but only a short way down. In the short-styled form the anthers are placed in the mouth of the corolla above the stigma, which occupies the same position as the anthers in the other form, being seated only a short way down the tube. Therefore the pistil of the long-styled form does not exceed in length that of the short-styled in nearly so great a degree as in many other Rubiaceae. Nevertheless there is a considerable difference in the size of the pollen grains

[25] Fritz Müller gives another instance of the want of absolute perfection in the flowers of another member of the Rubiaceae, namely, *Posoqueria fragrans*, which is adapted in a most wonderful manner for cross-fertilization by the agency of moths. (See *Bot. Zeitung*, 1866, No. 17.) In accordance with the nocturnal habits of these insects, most of the flowers open only during the night; but some open in the day, and the pollen of such flowers is robbed, as Fritz Müller has often seen by humble-bees and other insects, without any benefit being thus conferred on the plant.

in the two forms; for, as Fritz Müller informs me, those of the short-styled are to those of the long-styled as 100 to 75 in diameter.

HOUSTONIA COERULEA (RUBIACEAE)

Professor Asa Gray has been so kind as to send me an abstract of some observations made by Dr Rothrock on this plant. The pistil is exserted in the one form and the stamens in the other, as has long been observed. The stigmas of the long-styled form are shorter, stouter, and far more hispid than in the other form. The stigmatic hairs or papillae on the former are 0·04 mm, and on the latter only 0·023 mm in length. In the short-styled form the anthers are larger, and the pollen grains, when distended with water, are to those from the long-styled form as 100 to 72 in diameter.

Selected capsules from some long-styled plants growing in the Botanic Gardens at Cambridge, U.S., near where plants of the other form grew, contained on an average 13 seeds; but these plants must have been subjected to unfavourable conditions, for some long-styled plants in a state of nature yielded an average of 21·5 seeds per capsule. Some short-styled plants, which had been planted by themselves in the Botanic Gardens, where it was not likely that they would have been visited by insects that had previously visited long-styled plants, produced capsules, eleven of which were wholly sterile, but one contained 4, and another 8 seeds. So that the short-styled form seems to be very sterile with its own pollen. Professor Asa Gray informs me that the other North American species of this genus are likewise heterostyled.

OLDENLANDIA [SP.?] (RUBIACEAE)

Mr J. Scott sent me from India dried flowers of a heterostyled species of this genus, which is closely allied to the last. / The pistil in the long-styled flowers is longer by about a quarter of its length, and the stamens shorter in about the same proportion, than the corresponding organs in the short-styled flowers. In the latter the anthers are longer, and the divergent stigmas decidedly longer and apparently thinner than in the long-styled form. Owing to the state of the specimens, I could not decide whether the stigmatic papillae were longer in the one form than in the other. The pollen grains, distended with water, from the short-styled flowers were to those from the long-styled as 100 to 78 in diameter, as deduced from the mean of ten measurements of each kind.

HEDYOTIS [SP.?] (RUBIACEAE)

Fritz Müller sent me from St Catharina, in Brazil, dried flowers of a small delicate species, which grows on wet sand near the edges of fresh-water pools. In the long-styled form the stigma projects above the corolla, and stands on a level with the projecting anthers of the short-styled form; but in the latter the stigmas stand rather beneath the level of the anthers in the other or long-styled form, these being enclosed within the tube of the corolla. The pistil of the long-styled form is nearly thrice as long as that of the short-styled, or, speaking strictly, as 100 to 39; and the papillae on the stigma of the former are broader, in the ratio of 4 to 3, but whether longer than those of the short-styled, I could not decide. In the short-styled form, the anthers are rather larger, and the pollen grains are to those from the long-styled flowers as 100 to 88 in diameter. Fritz Müller sent me a second, small-sized species, which is likewise heterostyled.

COCCOCYPSELUM [SP.?] (RUBIACEAE)

Fritz Müller also sent me dried flowers of this plant from St Catharina, in Brazil. The exserted stigma of the long-styled form stands a little above the level of the exserted anthers of the short-styled form; and the enclosed stigma of the latter also stands a little above the level of the enclosed anthers in the long-styled form. The pistil of the long-styled is about twice as long as that of the short-styled, with its two stigmas considerably longer, more divergent, and more curled. Fritz Müller informs / me that he could detect no difference in the size of the pollen grains in the two forms. Nevertheless, there can be no doubt that this plant is heterostyled.

LIPOSTOMA [SP.?] (RUBIACEAE)

Dried flowers of this plant, which grows in small wet ditches in St Catharina, in Brazil, were likewise sent me by Fritz Müller. In the long-styled form the exserted stigma stands rather above the level of the exserted anthers of the other form; whilst in the short-styled form it stands on a level with the anthers of the other form. So that the want of strict correspondence in height between the stigmas and anthers in the two forms is reversed, compared with what occurs in Hedyotis. The long-styled pistil is to that of the short-styled as 100 to 36 in length; and its divergent stigmas are longer by fully one-third of their own length than those of the short-styled form. In the latter the anthers are a little larger, and the pollen grains are as 100 to 80 in diameter, compared with those from the long-styled form.

CINCHONA MICRANTHA (RUBIACEAE)

Dried specimens of both forms of this plant were sent me from Kew.[26] In the long-styled form the apex of the stigma stands just beneath the bases of the hairy lobes of the corolla; whilst the summits of the anthers are seated about halfway down the tube. The pistil is in length as 100 to 38 to that of the short-styled form. In the latter the anthers occupy the same position as the stigma of the other form, and they are considerably longer than those of the long-styled

[26] My attention was called to this plant by a drawing copied from Howard's *Quinologia*, Tab. 3, given by Mr Markham in his *Travels in Peru*, p. 539.

form. As the summit of the stigma in the short-styled form stands beneath the bases of the anthers, which are seated halfway down the corolla, the style has been extremely shortened in this form; its length to that of the long-styled being, in the specimens examined, only as 5·3 to 100! The stigma, also, in the short-styled form is very much shorter than that in the long-styled, in the ratio of 57 to 100. The pollen grains from the short-styled / flowers, after having been soaked in water, were rather larger – in about the ratio of 100 to 91 – than those from the long-styled flowers, and they were more triangular, with the angles more prominent. As all the grains from the short-styled flowers were thus characterized, and as they had been left in water for three days, I am convinced that this difference in shape in the two sets of grains cannot be accounted for by unequal distension with water.

Besides the several Rubiaceous genera already mentioned, Fritz Müller informs me that two or three species of Psychotria and *Rudgea eriantha*, natives of St Catharina, in Brazil, are heterostyled, as is *Manettia bicolor*. I may add that I formerly fertilized with their own pollen several flowers on a plant of this latter species in my hothouse, but they did not set a single fruit. From Wight and Arnott's description, there seems to be little doubt that Knoxia in India is heterostyled; and Asa Gray is convinced that this is the case with Diodia and Spermacoce in the United States. Lastly, from Mr W. W. Bailey's description,[27] it appears that the Mexican *Bouvardia leiantha* is heterostyled.

Altogether we now know of 17 heterostyled genera in the great family of the Rubiaceae, though more information is necessary with respect to some of them, more especially those mentioned in the last paragraph, before we can feel absolutely safe. In the *Genera Plantarum*, by Bentham and Hooker, the Rubiaceae are divided into 25 tribes, containing 337 genera; and it deserves notice that the genera now known to be heterostyled are not grouped in one or two of these tribes, but are distributed in no less than eight of them. From this fact we may infer that most of the genera have acquired their heterostyled structure independently of one another; that is, they have not inherited this structure from some one or even two or three progenitors in common. It further / deserves notice that in the homostyled genera, as I am informed by Professor Asa Gray, the stamens are either exserted or are included within the tube of the corolla, in a nearly constant manner; so that this character, which is not even of specific value in the heterostyled species, is often of generic value in other members of the family. /

[27] *Bull. of the Torrey Bot. Club*, 1876, p. 106.

CHAPTER IV

HETEROSTYLED TRIMORPHIC PLANTS

Lythrum salicaria – Description of the three forms – Their power and complex manner of fertilizing one another – Eighteen different unions possible – Mid-styled form eminently feminine in nature – *Lythrum Graefferi* likewise trimorphic – *L. thymifolia* dimorphic – *L. hyssopifolia* homostyled – *Nesaea verticillata* trimorphic – Lagerstroemia, nature doubtful – Oxalis, trimorphic species of – *O. Valdiviana* – *O. Regnelli*, the illegitimate unions quite barren – *O. speciosa* – *O. sensitiva* – Homostyled species of Oxalis – Pontederia, the one monocotyledonous genus known to include heterostyled species.

In the previous chapters various heterostyled dimorphic plants have been described, and now we come to heterostyled trimorphic plants, or those which present three forms. These have been observed in three families, and consist of species of Lythrum and of the allied genus Nesaea, of Oxalis and Pontederia. In their manner of fertilization these plants offer a more remarkable case than can be found in any other plant or animal.

Lythrum salicaria. The pistil in each form differs from that in either of the other forms, and in each there are two sets of stamens different in appearance and function. But one set of stamens in each form corresponds with a set in one of the other two forms. Altogether this one species includes three females or female organs and three sets of male organs, all as distinct from one another as if they belonged to different species; and if smaller functional differences / are considered, there are five distinct sets of males. Two of the three hermaphrodites must co-exist, and pollen must be carried by insects reciprocally from one to the other, in order that either of the two should be fully fertile; but unless all three forms co-exist, two sets of stamens will be wasted, and the organization of the species, as a whole, will be incomplete. On the other hand, when all three hermaphrodites co-exist, and pollen is carried from one to the other, the scheme is

98

perfect; there is no waste of pollen and no false co-adaptation. In short, nature has ordained a most complex marriage-arrangement, namely a triple union between three hermaphrodites – each hermaphrodite being in its female organ quite distinct from the other two hermaphrodites and partially distinct in its male organs, and each furnished with two sets of males.

The three forms may be conveniently called, from the unequal lengths of their pistils, the *long-styled, mid-styled*, and short-styled. The stamens also are of unequal length, and these may be called the *longest, mid-length*, and *shortest*. Two sets of stamens of different length are found in each form. The existence of the three forms was first observed by Vaucher,[1] and subsequently more carefully by Wirtgen; but these botanists, not being guided by any theory or even suspicion of their functional differences, did not perceive some of the most curious points of difference in their structure. I will first briefly describe the three forms by the aid of the accompanying diagram, which shows the flowers, six times magnified, in their natural position, with their petals and calyx on the near side removed. /

Long-styled form. This form can be at once recognized by the length of the pistil, which is (including the ovarium) fully one-third longer than that of the mid-styled, and more than thrice as long as that of the short-styled form. It is so disproportionately long, that it projects in the bud through the folded petals. It stands out considerably beyond the mid-length stamens; its terminal portion depends a little, but the stigma itself is slightly upturned. The globular stigma is considerably larger than that of the other two forms, with the papillae on its surface generally longer. The six mid-length stamens project about two-thirds the length of the pistil, and correspond in length with the pistil of the mid-styled form. Such correspondence in this and the two following forms is generally very close; the difference, where there is any, being usually in a slight excess of length in the stamens. The six shortest stamens lie concealed within the calyx; their ends are turned up, and they are graduated in length, so as to form a double row. The anthers of these stamens are smaller than those of the mid-length ones. The pollen is of the same yellow colour in both sets. H. Müller[2] measured

[1] *Hist. Phys. des Plantes d'Europe*, vol. ii, 1841, p. 371. Wirtgen, 'Ueber *Lythrum salicaria* und dessen Formen', *Verhand. des naturhist. Vereins für preuss. Rheinl.*, 5. Jahrgang, 1848, p. 7.
[2] *Die Befruchtung der Blumen*, 1873, p. 193.

99

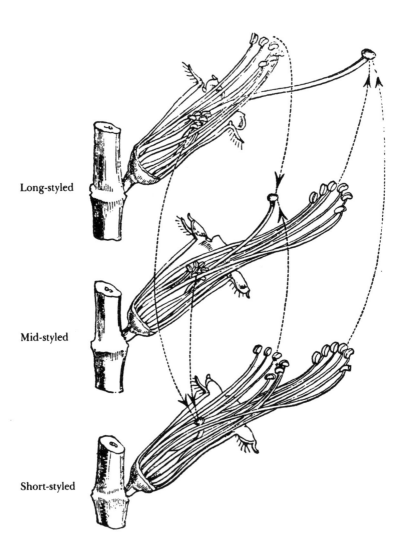

Fig. 10 Diagram of the flowers of the three forms of *Lythrum salicaria*, in their natural position, with the petals and calyx removed on the near side: enlarged six times.
The dotted lines with the arrows show the directions in which pollen must be carried to each stigma to ensure full fertility.

the pollen grain in all three forms, and his measurements are evidently more trustworthy than those which I formerly made, so I will give them. The numbers refer to divisions of the micrometer equalling ⅓₀₀ mm. The grains, distended with water, from the mid-length stamens are 7–7½, and those from the shortest stamens 6–6½ in diameter, or as 100 to 86. The capsules of this form contain on an average 93 seeds; how this average was obtained will presently be explained. As these seeds, when cleaned, seemed larger than those from the mid-styled / or short-styled forms, 100 of them were placed in a good balance, and by the double method of weighing were found to equal 121 seeds of the mid-styled or 142 of the short-styled; so that five long-styled seeds very nearly equal six mid-styled or seven short-styled seeds.

Mid-styled form. The pistil occupies the position represented in the diagram, with its extremity considerably upturned, but to a variable degree; the stigma is seated between the anthers of the longest and the shortest stamens. The six longest stamens correspond in length with the pistil of the long-styled form; their filaments are coloured bright pink; the anthers are dark-coloured, but from containing bright-green pollen and from their early dehiscence they appear emerald-green. Hence in general appearance these stamens are remarkably dissimilar from the mid-length stamens of the long-styled form. The six shortest stamens are enclosed within the calyx, and resemble in all respects the shortest stamens of the long-styled form; both these sets correspond in length with the short pistil of the short-styled form. The green pollen grains of the longest stamens are 9–10 in diameter, whilst the yellow grains from the shortest stamens are only 6; or as 100 to 63. But the pollen grains from different plants appeared to me, in this case and others, to be in some degree variable in size. The capsules contain on an average 130 seeds; but perhaps, as we shall see, this is rather too high an average. The seeds themselves, as before remarked, are smaller than those of the long-styled form.

Short-styled form. The pistil is here very short, not one-third of the length of that of the long-styled form. It is enclosed within the calyx, which, differently from that in the other two forms, does not enclose any anthers. / The end of the pistil is generally bent upwards at right angles. The six longest stamens, with their pink filaments and green pollen, resemble the corresponding stamens of the mid-styled form. But according to H. Müller, their pollen grains are a little larger, viz.

9½–10½, instead of 9–10 in diameter. The six mid-length stamens, with their uncoloured filaments and yellow pollen, resemble in the size of their pollen grains and in all other respects the corresponding stamens of the long-styled form. The difference in diameter between the grains from the two sets of anthers in the short-styled form is as 100 to 73. The capsules contain fewer seeds on an average than those of either of the preceding forms, namely 83·5; and the seeds are considerably smaller. In this latter respect, but not in number, there is a gradation parallel to that in the length of the pistil, the long-styled having the largest seeds, the mid-styled the next in size, and the short-styled the smallest.

We thus see that this plant exists under three female forms, which differ in the length and curvature of the style, in the size and state of the stigma, and in the number and size of the seed. There are altogether thirty-six males or stamens, and these can be divided into three sets of a dozen each, differing from one another in length, curvature, and colour of the filaments – in the size of the anthers, and especially in the colour and diameter of the pollen grains. Each form bears half-a-dozen of one kind of stamens and half-a-dozen of another kind, but not all three kinds. The three kinds of stamens correspond in length with the three pistils; the correspondence is always between half of the stamens in two of the forms with the pistil of the third form. The following table of the diameters of the pollen grains, after immersion in water, from / both sets of stamens in all three forms is copied from H. Müller; they are arranged in the order of their size:

Pollen grains from longest stamens of short-styled form	9½ to 10½
Pollen grains from longest stamens of mid-styled form	9 to 10
Pollen grains from mid-length stamens of long-styled form	7 to 7½
Pollen grains from mid-length stamens of short-styled form	7 to 7½
Pollen grains from shortest stamens of long-styled form	6 to 6½
Pollen grains from shortest stamens of mid-styled form	6 to 6

We here see that the largest pollen grains come from the longest stamens, and the least from the shortest; the extreme difference in diameter between them being as 100 to 60.

The average number of seeds in the three forms was ascertained by counting them in eight fine selected capsules taken from plants growing wild, and the result was, as we have seen, for the long-styled (neglecting decimals) 93, mid-styled 130, and short-styled 83. I should not have trusted in these ratios had I not possessed a number of plants in my garden which, owing to their youth, did not yield the full complement of seed, but were of the same age and grew under the same conditions, and were freely visited by bees. I took six fine capsules from each, and found the average to be for the long-styled 80, for the mid-styled 97, and for the short-styled 61. Lastly, legitimate unions effected by me between the three forms gave, as may be seen in the following tables, for the long-styled an average of 90 seeds, for the mid-styled 117, and for the short-styled 71. So that we have good concurrent evidence of a difference in the average production of seed by the three forms. To show that the unions effected by me often produced their full effect and may be trusted, I may state that one mid-styled capsule yielded 151 good seeds, which is the same number as in the finest wild capsules which I / examined. Some artificially fertilized short- and long-styled capsules produced a greater number of seeds than was ever observed by me in wild plants of the same forms, but then I did not examine many of the latter. This plant, I may add, offers a remarkable instance, how profoundly ignorant we are of the life-conditions of a species. Naturally it grows 'in wet ditches, watery places, and especially on the banks of streams', and though it produces so many minute seeds, it never spreads on the adjoining land; yet, when planted in my garden, on clayey soil lying over chalk, and which is so dry that a rush cannot be found, it thrives luxuriantly, grows to above 6 feet in height, produces self-sown seedlings, and (which is a severer test) is as fertile as in a state of nature. Nevertheless it would be almost a miracle to find this plant growing spontaneously on such land as that in my garden.

According to Vaucher and Wirtgen, the three forms co-exist in all parts of Europe. Some friends gathered for me in North Wales a number of twigs from separate plants growing near one another, and classified them. My son did the same in Hampshire, and here is the result:

TABLE 22

—	Long-styled	Mid-styled	Short-styled	Total
North Wales	95	97	72	264
Hampshire	53	38	38	129
Total	148	135	110	393

If twice or thrice the number had been collected, the three forms would probably have been found nearly equal; I infer this from considering the above figures, and from my son telling me that if he had / collected in another spot he felt sure that the mid-styled plants would have been in excess. I several times sowed small parcels of seed, and raised all three forms; but I neglected to record the parent form, excepting in one instance, in which I raised from short-styled seed twelve plants, of which only one turned out long-styled, four mid-styled, and seven short-styled.

Two plants of each form were protected from the access of insects during two successive years, and in the autumn they yielded very few capsules and presented a remarkable contrast with the adjoining uncovered plants, which were densely covered with capsules. In 1863 a protected long-styled plant produced only five poor capsules; two mid-styled plants produced together the same number; and two short-styled plants only a single one. These capsules contained very few seeds; yet the plants were fully productive when artificially fertilized under the net. In a state of nature the flowers are incessantly visited for their nectar by hive- and other bees, various Diptera and Lepidoptera.[3] The nectar is secreted all round the base of the ovarium; but a passage is formed along the upper and inner side of the flower by the lateral deflection (not represented in the diagram) of the basal portions of the filaments; so that insects invariably alight on the projecting stamens and pistil, and insert their proboscides along the upper and inner margin of the corolla. We can now see why the ends of the stamens with their anthers, and the ends of the pistils with their stigmas, / are a little upturned, so that they may be brushed by the lower hairy surfaces of the insects' bodies. The shortest stamens which lie enclosed within the calyx of the long- and mid-styled forms can be touched only by the proboscis and narrow chin of a bee; hence they have their ends more upturned, and they are graduated in length, so as to fall into a narrow file, sure to be raked by the thin intruding proboscis. The anthers of the longer stamens stand laterally farther apart and are more nearly on the same level, for they have to brush against the whole breadth of the insect's body. In very many other flowers the pistil, or the stamens, or both, are rectangularly bent to one side of the flower. This bending may be permanent, as with Lythrum and many others, or may be effected, as in *Dictamnus fraxinella* and others, by a temporary movement, which occurs in the case of the stamens, when

---

[3] H. Müller gives a list of the species, *Die Befruchtung der Blumen*, p. 196. It appears that one bee, the *Cilissa melanura*, almost confines its visits to this plant.

the anthers dehisce, and in the case of the pistil when the stigma is mature; but these two movements do not always take place simultaneously in the same flower. Now I have found no exception to the rule, that when the stamens and pistil are bent, they bend to that side of the flower which secretes nectar, even though there be a rudimentary nectary of large size on the opposite side, as in some species of Corydalis. When nectar is secreted on all sides, they bend to that side where the structure of the flower allows the easiest access to it, as in Lythrum, various Papilionaceae, and others. The rule consequently is, that when the pistils and stamens are curved or bent, the stigma and anthers are thus brought into the pathway leading to the nectary. There are a few cases which seem to be exceptions to this rule, but they are not so in truth; for instance, in the Gloriosa lily, the stigma of the grotesque and rectangularly bent pistil / is brought, not into any pathway from the outside towards the nectar-secreting recesses of the flower, but into the circular route which insects follow in proceeding from one nectary to the other. In *Scrophularia aquatica* the pistil is bent downwards from the mouth of the corolla, but it thus strikes the pollen-dusted breast of the wasps which habitually visit these ill-scented flowers. In all these cases we see the supreme dominating power of insects on the structure of flowers, especially of those which have irregular corollas. Flowers which are fertilized by the wind must of course be excepted; but I do not know of a single instance of an irregular flower which is thus fertilized.

Another point deserves notice. In each of the three forms two sets of stamens correspond in length with the pistils in the other two forms. When bees suck the flowers, the anthers of the longest stamens, bearing the green pollen, are rubbed against the abdomen and the inner sides of the hind legs, as is likewise the stigma of the long-styled form. The anthers of the mid-length stamens and the stigma of the mid-styled form are rubbed against the under side of the thorax and between the front pair of legs. And, lastly, the anthers of the shortest stamens and the stigma of the short-styled form are rubbed against the proboscis and chin; for the bees in sucking the flowers insert only the front part of their heads into the flower. On catching bees, I observed much green pollen on the inner sides of the hind legs and on the abdomen, and much yellow pollen on the under side of the thorax. There was also pollen on the chin, and, it may be presumed, on the proboscis, but this was difficult to observe. I had, however, independent proof that pollen is carried on the proboscis; for a small branch of a protected short-styled plant (which produced spontaneously only two / capsules) was accidentally left during several days pressing against the net, and bees were seen inserting their proboscides through the meshes, and in consequence numerous capsules were formed on this one small branch. From these several facts it follows that insects will generally carry the pollen of each form from the stamens to the pistil of corresponding length; and we shall presently see the importance of this adaptation. It must not, however, be supposed that the bees do not get more or less dusted all over with the several kinds of pollen; for this could be seen to occur with the green pollen from the longest stamens. Moreover a case will presently be given of a long-styled plant producing an abundance of capsules, though growing quite by itself, and the flowers must have been fertilized by their own two kinds of pollen; but these capsules contained a very poor average of seed. Hence insects, and chiefly bees, act both as general carriers of pollen, and as special carriers of the right sort.

Wirtgen remarks[4] on the variability of this plant in the branching of the stem, in the length of the bracteae, size of the petals, and in several other characters. The plants which grew in my garden had their leaves, which differed much in shape, arranged oppositely, alternately, or in whorls of three. In this latter case the stems were hexagonal; those of the other plants being quadrangular. But we are concerned chiefly, with the reproductive organs: the upward bending of the pistil is variable, and especially in the short-styled form, in which it is sometimes straight, sometimes slightly curved, but generally bent at right angles. The stigma of the long-styled pistil frequently has / longer papillae or is rougher than that of the mid-styled, and the latter than that of the short-styled; but this character, though fixed and uniform in the two forms of *Primula veris*, etc., is here variable, for I have seen mid-styled stigmas rougher than those of the long-styled.[5] The degree to which the longest and mid-length stamens are graduated in length and have their ends upturned is variable; sometimes all are equally long. The colour of the green pollen in the longest stamens is variable, being sometimes pale greenish-yellow; in one short-styled plant it was

[4] *Verhand. des naturhist. Vereins, für Pr. Rheinl.*, 5. Jahrgang, 1848, pp. 11, 13.

[5] The plants which I observed grew in my garden, and probably varied rather more than those growing in a state of nature. H. Müller has described the stigmas of all three forms with great care, and he appears to have found the stigmatic papillae differing constantly in length and structure in the three forms, being longest in the long-styled form.

almost white. The grains vary a little in size: I examined one short-styled plant with the grains above the average size; and I have seen a long-styled plant with the grains from the mid-length and shortest anthers of the same size. We here see great variability in many important characters; and if any of these variations were of service to the plant, or were correlated with useful functional differences, the species is in that state in which Natural Selection might readily do much for its modification.

On the power of mutual fertilization between the three forms

Nothing shows more clearly the extraordinary complexity of the reproductive system of this plant, than the necessity of making eighteen distinct unions in order to ascertain the relative fertilizing power of the / three forms. Thus the long-styled form has to be fertilized with pollen from its own two kinds of anthers, from the two in the mid-styled, and from the two in the short-styled form. The same process has to be repeated with the mid-styled and short-styled forms. It might have been thought sufficient to have tried on each stigma the green pollen, for instance, from either the mid- or short-styled longest stamens, and not from both; but the result proves that this would have been insufficient, and that it was necessary to try all six kinds of pollen on each stigma. As in fertilizing flowers there will always be some failures, it would have been advisable to have repeated each of the eighteen unions a score of times; but the labour would have been too great; as it was, I made 223 unions, i.e. on an average I fertilized above a dozen flowers in the eighteen different methods. Each flower was castrated; the adjoining buds had to be removed, so that the flowers might be safely marked with thread, wool, etc.; and after each fertilization the stigma was examined with a lens to see that there was sufficient pollen on it. Plants of all three forms were protected during two years by large nets on a framework; two plants were used during one or both years, in order to avoid any individual peculiarity in a particular plant. As soon as the flowers had withered, the nets were removed; and in the autumn the capsules were daily inspected and gathered, the ripe seeds being counted under the microscope. I have given these details that confidence may be placed in the following tables, and as some excuse for two blunders which, I believe were made. These blunders are referred to, with their probable cause, in two footnotes to the tables. The erroneous numbers, however, are entered in the tables, that it may not be supposed / that I have in any one instance tampered with the results.

A few words explanatory of the three tables must be given. Each is devoted to one of the three forms, and is divided into six compartments. The two upper ones in each table show the number of good seeds resulting from the application to the stigma of pollen from the two sets of stamens which correspond in length with the pistil of that form, and which are borne by the other two forms. Such unions are of a legitimate nature. The two next lower compartments show the result of the application of pollen from the two sets of stamens not corresponding in length with the pistil, and which are borne by the other two forms. These unions are illegitimate. The two lowest compartments show the result of the application of each form's own two kinds of pollen from the two sets of stamens belonging to the same form, and which do not equal the pistil in length. These unions are likewise illegitimate. The term own-form pollen here used does not mean pollen from the flower to be fertilized – for this was never used – but from another flower on the same plant, or more commonly from a distinct plant of the same form. The figure (o) means that no capsule was produced, or if a capsule was produced that it contained no good

TABLE 23   Long-styled form

I

Legitimate union

13 flowers fertilized by the longest stamens of the mid-styled. These stamens equal in length the pistil of the long-styled.

Product of good seed in each capsule

36	53
81	o
o	o
o	o
o	o
—	o
45	
41	

38 per cent of these flowers yielded capsules. Each capsule contained, on an average, 51·2 seeds.

II

Legitimate union

13 flowers fertilized by the longest stamens of the short-styled. These stamens equal in length the pistil of the long-styled.

Product of good seed in each capsule

159	104
43	119
96 poor seed	96
103	99
o	131
o	116
114	

83 per cent of these flowers yielded capsules. Each capsule contained, on an average, 107·3 seeds.

TABLE 23   *Long-styled form – continued*

III		IV	
Illegitimate union		*Illegitimate union*	
14 flowers fertilized by the shortest stamens of the mid-styled		12 flowers fertilized by the mid-length stamens of the short-styled	
3	0	20	0
0	0	0	0
0	0	0	0
0	0	0	0
0	0	—	0
—	0	0	0
0	0	0	
0			
Too sterile for any average		Too sterile for any average	

V		VI	
Illegitimate union		*Illegitimate union*	
15 flowers fertilized by own-form mid-length stamens		15 flowers fertilized by own-form shortest stamens	
2	—	4	—
10	0	8	0
23	0	4	0
0	0	0	0
0	0	0	0
0	0	0	0
0	0	0	0
Too sterile for any average		Too sterile for any average	

seed. In some part of each row of figures in each compartment, a short horizontal line may be seen; the unions above this line were made in 1862, and below it in 1863. It is of importance to observe this, as it shows that the same general result was obtained during two successive years; but more especially because 1863 was a very hot and dry season, and the plants had occasionally to be watered. This did not prevent the full complement of seeds being produced from the more fertile unions; but it rendered the less fertile / ones even more sterile than they otherwise would have been. I have seen striking instances of this fact in making illegitimate and legitimate unions with Primula; and it is well known that the conditions of life must be highly favourable to give any chance of success in producing hybrids between species which are crossed with difficulty. /

Besides the above experiments, I fertilized a considerable number of long-styled flowers with pollen, taken by a camel's-hair brush, from both the mid-length and shortest stamens of their own form: only 5 capsules were produced, and these yielded on an average 14·5 seeds. In 1863 I tried a much better experiment: a long-styled plant was grown by itself, miles away from any other plant, so that the flowers could have received only their own two kinds of pollen. The flowers were incessantly visited by bees, and their stigmas must have received successive applications of pollen on the most favourable days and at the most favourable hours: all who have crossed plants know that this highly favours fertilization. This plant produced an abundant crop of capsules; I took by chance 20 capsules, and these contained seeds in number as follows:

20	20	35	21	19
26	24	12	23	10
7	30	27	29	13
20	12	29	19	35

/

TABLE 24   *Mid-styled form*

I	II
Legitimate union	*Legitimate union*
12 flowers fertilized by the mid-length stamens of the long-styled. These stamens equal in length the pistil of the mid-styled	12 flowers fertilized by the mid-length stamens of the short-styled. These stamens equal in length the pistil of the mid-styled
Product of good seed in each capsule	Product of good seed in each capsule

138	122	112	109
149	50	130	143
147	151	143	124
109	119	100	145
133	138	33	12
144	0	───	141
───		104	

92 per cent of the flowers (probably 100 per cent) yielded capsules. Each capsule contained, on an average, 127·3 seeds.

100 per cent of the flowers yielded capsules. Each capsule contained, on an average, 108·0 seeds; or, excluding capsules with less than 20 seeds, the average is 116·7 seeds./

TABLE 24 *Mid-styled form — continued*

III

Illegitimate union

13 flowers fertilized by the shortest stamens of the long-styled

83	12
0	19
0	85 seeds small and poor
—	0
44	0
44	0
45	0

54 per cent of the flowers yielded capsules. Each capsule contained, on an average, 47·4 seeds; or, excluding capsules with less than 20 seeds, the average is 60·2 seeds.

IV

Illegitimate union

15 flowers fertilized by the longest stamens of the short-styled

130	86
115	113
14	29
6	17
2	113
9	79
—	128
132	0

93 per cent of the flowers yielded capsules. Each capsule contained, on an average, 69·5 seeds; or, excluding capsules with less than 20 seeds, the average is 102·8.

V

Illegitimate union

12 flowers fertilized by own-form longest stamens

92	0
9	0
63	0
——	0
136?[6]	0
0	0
0	

Excluding the capsule with 136 seeds, 25 per cent of the flowers yielded capsules, and each capsule contained, on an average, 54·6 seeds; or, excluding capsules with less than 20 seeds, the average is 77·5.

VI

Illegitimate union

12 flowers fertilized by own-form shortest stamens

0	0
0	0
0	0
–	0
0	0
0	0
0	

Not one flower yielded a capsule.

[6] I have hardly a doubt that this result of 136 seeds in compartment V was due to a gross error. The flowers to be fertilized by their own longest stamens were first marked by 'white thread', and those by the mid-length stamens of the long-styled form by 'white silk'; a flower fertilized in the later manner would have yielded about 136 seeds, and it may be observed that one such pod is missing, viz. at the bottom of compartment I. Therefore I have hardly any doubt that I fertilized a flower marked with 'white thread' as if it had been marked with 'white silk'. With respect to the

This gives an average of 21·5 seeds per capsule. As we know that the long-styled form, when standing near plants of the other two forms and fertilized by insects, produces on an average 93 seeds per capsule, we see that this form, fertilized by its own two pollens, yields only between one-fourth and one-fifth of the full number of seed. I have spoken as if the plant had received both its own kinds of pollen, and this is, of course, possible; but, from the enclosed position of the shortest stamens, it is much more probable that the stigma received exclusively pollen from the mid-length stamens; and this, as may be seen in compartment V in Table 23, is the more fertile of the two self-unions. /

Besides the experiments in Table 24, I fertilized a considerable number of mid-styled flowers with pollen, taken by a camel's-hair brush, from both the longest and shortest stamens of their own form: only 5 capsules were produced, and these yielded on an average 11·0 seeds.

TABLE 25   *Short-styled form*

I		II	
Legitimate union		*Legitimate union*	
12 flowers fertilized by the shortest stamens of the long-styled. These stamens equal in length the pistil of the short-styled		13 flowers fertilized by the shortest stamens of the mid-styled. These stamens equal in length the pistil of the short-styled	
69	56	93	69
61	88	77	69
88	112	48	53
66	111	43	9
o	62	o	o
o	100	o	o
—		—	o

83 per cent of the flowers yielded capsules. Each capsule contained, on an average, 81·3 seeds.

61 per cent of the flowers yielded capsules. Each capsule contained, on an average, 64·6 seeds.

---

capsule which yielded 92 seeds, in the same column with that which yielded 136, I do not know what to think. I endeavoured to prevent pollen dropping from an upper to a lower flower, and I tried to remember to wipe the pincers carefully after each fertilization; but in making eighteen different unions, sometimes on windy days, and pestered by bees and flies buzzing about, some few errors could hardly be avoided. One day I had to keep a third man by me all the time to prevent the bees visiting the uncovered plants, for in a few seconds' time they might have done

TABLE 25   *Short-styled form – continued*

III		IV	
Illegitimate union		*Illegitimate union*	
10 flowers fertilized by the mid-length stamens of the long-styled		10 flowers fertilized by the longest stamens of the mid-styled	
o	14	o	o
o	o	o	o
o	o	o	o
o	o	o	o
—	o	—	o
23		o	
Too sterile for any average		Too sterile for any average	

V		VI	
Illegitimate union		*Illegitimate union*	
10 flowers fertilized by own-form longest stamens		10 flowers fertilized by own-form mid-length stamens	
o	o	64?[7]	o
o	o	o	o
o	o	o	o
—	o	—	o
o	o	21	o
o		9	
Too sterile for any average		Too sterile for any average	

Besides the experiments in the table, I fertilized a number of flowers without particular care with their own two kinds of pollen, but they did not produce a single capsule.

Summary of the results

Long-styled form. Twenty-six flowers fertilized legitimately by the stamens of corresponding length, borne by the mid- and short-styled

---

irreparable mischief. It was also extremely difficult to exclude minute Diptera from the net. In 1862 I made the great mistake of placing a mid-styled and long-styled under the same huge net: in 1863 I avoided this error.

[7] I suspect that by mistake I fertilized this flower in compartment VI with pollen from the shortest stamens of the long-styled form and it would then have yielded about 64 seeds. Flowers to be thus fertilized were marked with black silk; those with pollen from the mid-length stamens of the short-styled with black thread: and thus probably the mistake arose.

forms, yielded 61·5 per cent of capsules, which contained on an average 89·7 seeds.

Twenty-six long-styled flowers fertilized illegitimately by the other stamens of the mid- and short-styled forms yielded only two very poor capsules.

Thirty long-styled flowers fertilized illegitimately by their own-form two sets of stamens yielded only eight very poor capsules; but long-styled flowers fertilized / by bees with pollen from their own stamens produced numerous capsules containing on an average 21·5 seeds.

Mid-styled form. Twenty-four flowers legitimately fertilized by the stamens of corresponding length, borne by the long- and short-styled forms, yielded 96 (probably 100) per cent of capsules, which contained (excluding one capsule with 12 seeds) on an average 117·2 seeds.

Fifteen mid-styled flowers fertilized illegitimately by the longest stamens of the short-styled form yielded 93 per cent of capsules, which (excluding four capsules with less than 20 seeds) contained on an average 102·8 seeds.

Thirteen mid-styled flowers fertilized illegitimately by the mid-length stamens of the long-styled form yielded 54 per cent of capsules, which (excluding one with 19 seeds) contained on an average 60·2 seeds.

Twelve mid-styled flowers fertilized illegitimately by their own-form longest stamens yielded 25 per cent of capsules, which (excluding one with 9 seeds) contained on an average 77·5 seeds.

Twelve mid-styled flowers fertilized illegitimately by their own-form shortest stamens yielded not a single capsule.

Short-styled form. Twenty-five flowers fertilized legitimately by the stamens of corresponding length, borne by the long- and mid-styled forms, yielded 72 per cent of capsules, which (excluding one capsule with only 9 seeds) contained on an average 70·8 seeds.

Twenty short-styled flowers fertilized illegitimately by the other stamens of the long- and mid-styled forms yielded only two very poor capsules.

Twenty short-styled flowers fertilized illegitimately / by their own stamens yielded only two poor (or perhaps three) capsules.

If we take all six legitimate unions together, and all twelve illegitimate unions together, we get the following results:

TABLE 26

Nature of union	Number of flowers fertilized	Number of capsules produced	Average number of seeds per capsule	Average number of seeds per flower fertilized
The six legitimate unions	75	56	96·29	71·89
The twelve illegitimate unions	146	36	44·72	11·03

Therefore the fertility of the legitimate unions to that of the illegitimate, as judged by the proportion of the fertilized flowers which yielded capsules, is as 100 to 33; and judged by the average number of seeds per capsules, as 100 to 46.

From this summary and the several foregoing tables we see that it is only pollen from the longest stamens which can fully fertilize the longest pistil; only that from the mid-length stamens, the mid-length pistil; and only that from the shortest stamens, the shortest pistil. And now we can comprehend the meaning of the almost exact correspondence in length between the pistil in each form and a set of six stamens in two of the other forms; for the stigma of each form is thus rubbed against that part of the insect's body which becomes charged with the proper pollen. It is also evident that the stigma of each form, fertilized in three different ways with pollen from the longest, mid-length, and shortest stamens, is acted on very differently, and conversely, that the pollen from / the twelve longest, twelve mid-length, and twelve shortest stamens acts very differently on each of the three stigmas; so that there are three sets of female and of male organs. Moreover, in most cases the six stamens of each set differ somewhat in their fertilizing power from the six corresponding ones in one of the other forms. We may further draw the remarkable conclusion that the greater the inequality in length between the pistil and the set of stamens, the pollen of which is employed for its fertilization, by so much is the sterility of the union increased. There are no exceptions to this rule. To understand what follows the reader should look to Tables 23, 24, and 25, and to the diagram Fig. 10, p. 100. In the long-styled form the shortest stamens obviously differ in length from the pistil to a greater degree than do the mid-length stamens; and the capsules produced by the use of pollen from the shortest stamens contain fewer seeds than those produced by the pollen from the mid-length stamens. The same result follows with the long-styled form, from the use of the pollen of the shortest stamens of the mid-styled form and of the mid-length stamens of the short-styled form. The same rule also holds good with the mid-styled and short-styled forms, when illegitimately fertilized with pollen from the stamens more or less unequal in length to their pistils. Certainly the difference in sterility in these several cases is slight; but, as far as we are enabled to judge, it always increases with the increasing inequality of length between the pistil and the stamens which are used in each case.

The correspondence in length between the pistil in each form and a set of stamens in the other two forms, is probably the direct result of adaptation, as it is of high service to the specie by leading to full and / legitimate fertilization. But the rule of the increased sterility of the illegitimate unions according to the greater inequality in length between the pistils and stamens employed for the union can be of no service. With some heterostyled dimorphic plants the difference of fertility between the two illegitimate unions appears at first sight to be related to the facility of self-fertilization; so that when from the position of the parts the liability in one form to self-fertilization is greater than in the other, a union of this kind has been checked by having been rendered the more sterile of the two. But this explanation does not apply to Lythrum; thus the stigma of the long-styled form is more liable to be illegitimately fertilized with pollen from its own mid-length stamens, or with pollen from the mid-length stamens of the short-styled form, than by its own shortest stamens or those of the mid-styled form; yet the two former unions, which it might have been expected would have been guarded against by increased sterility, are much less sterile than the other two unions which are much less likely to be effected. The same relation holds good even in a more striking manner with the mid-styled form, and with the short-styled form as far as the extreme sterility of all its legitimate unions allows of any comparison. We are led, therefore, to conclude that the rule of increased sterility, in accordance with increased inequality in length between the pistils and stamens, is a purposeless result, incidental on those changes through which the species has passed in acquiring certain characters fitted to ensure the legitimate fertilization of the three forms.

Another conclusion which may be drawn from Tables 23, 24, and 25, even from a glance at them, / is that the mid-styled form differs from both the others in its much higher capacity for fertilization in various ways. Not only did the twenty-four flowers legitimately fertilized by the stamens of corresponding lengths, all, or all but one, yield capsules rich in seed; but of the other four illegitimate unions, that by the longest stamens of the short-styled form, was highly fertile, though less so than the two legitimate unions, and that by the mid-length stamens of the long-styled form was fertile to a considerable degree; the remaining two illegitimate unions, namely, with this form's own pollen, were sterile, but in different degrees. So that the mid-styled form, when fertilized in the six different possible methods, evinces five grades of fertility. By comparing compartments III and VI in Table 24 we may see that the action of the pollen from the shortest stamens of the long-styled and mid-styled forms is widely different; in the one case above half the fertilized flowers yielded capsules containing a fair number of seeds; in the other case not one capsule was produced. So, again, the green, large-grained pollen from the longest stamens of the short-styled and mid-styled forms (in compartments IV and V) is widely different. In both these cases the difference in action is so plain that it cannot be mistaken, but it can be corroborated. If we look to Table 25 to the legitimate action of the shortest stamens of the long- and mid-styled forms on the short-styled form, we again see a similar but slighter difference, the pollen of the shortest stamens of the mid-styled form yielding a smaller average of seed during the two years of 1862 and 1863 than that from the shortest stamens of the long-styled form. Again, if we look to Table 23, to the legitimate action on the long-styled form of the green pollen of the two / sets of longest stamens, we shall find exactly the same result, viz. that the pollen from the longest stamens of the mid-styled form yielded during both years fewer seeds than that from the longest stamens of the short-styled form. Hence it is certain that the two kinds of pollen produced by the mid-styled form are less potent than the two similar kinds of pollen produced by the corresponding stamens of the other two forms.

In close connection with the lesser potency of the two kinds of pollen of the mid-styled form is the fact that, according to H. Müller, the grains of both are a little less in diameter than the corresponding grains produced by the other two forms. Thus the grains from the longest stamens of the mid-styled form are 9 to 10, whilst those from the corresponding stamens of the short-styled form are 9½ to 10½ in diameter. So, again, the grains from the shortest stamens of the mid-styled are 6, whilst those from the corresponding stamens of the long-styled are 6 to 6½ in diameter. It would thus appear as if the male organs of the mid-styled form, though not as yet rudimentary, were tending in this direction. On the other hand, the female organs of this form are in an eminently efficient state, for the naturally fertilized capsules yielded a considerably larger average number of seeds than those of the other two forms – almost every flower which was artificially fertilized in a legitimate manner produced a capsule – and most of the legitimate unions were highly productive. The mid-styled form thus appears to be highly feminine in nature; and although, as just remarked, it is impossible to consider its two well-developed sets of stamens which produce an abundance of pollen as being in a rudimentary condition, yet we can hardly avoid connecting as / balanced the higher efficiency of the female organs in this form with the lesser efficiency and lesser size of its two kinds of pollen grains. The whole case appears to me a very curious one.

It may be observed in Tables 23 to 25 that some of the illegitimate unions yielded during neither year a single seed; but, judging from the long-styled plants, it is probable, if such unions were to be effected repeatedly by the aid of insects under the most favourable conditions, some few seeds would be produced in every case. Anyhow, it is certain that in all twelve illegitimate unions the pollen tubes penetrated the stigma in the course of eighteen hours. At first I thought that two kinds of pollen placed together on the same stigma would perhaps yield more seed than one kind by itself; but we have seen that this is not so with each form's own two kinds of pollen; nor is it probable in any case, as I occasionally got, by the use of a single kind of pollen, fully as many seeds as a capsule naturally fertilized ever produces. Moreover the pollen from a single anther is far more than sufficient to fertilize fully a stigma; since, in this as with so many other plants, more than twelve times as much of each kind of pollen is produced as is necessary to ensure the full fertilization of each form. From the dusted condition of the bodies of the bees which I caught on the flowers, it is probable that pollen of various kinds is often deposited on all three stigmas; but from the facts already given with respect to the two forms of Primula, there can hardly be a doubt that pollen from the stamens of corresponding length placed on a stigma would be prepotent over any other kind of pollen and obliterate its effects – even if the latter had been placed on the stigma some hours previously. /

Finally, it has now been shown that *Lythrum salicaria* presents the extraordinary case of the same species bearing three females, different in structure and function, and three or even five sets (if minor differences are considered) of males; each set consisting of half-a-dozen, which likewise differ from one another in structure and function.

Lythrum Graefferi. I have examined numerous dried flowers of this species, each from a separate plant, sent me from Kew. Like *L. salicaria*, it is trimorphic, and the three forms apparently occur in about equal numbers. In the long-styled form the pistil projects about one-third of the length of the calyx beyond its mouth, and is therefore relatively much shorter than in *L. salicaria*; the globose and hirsute stigma is larger than that of the other two forms; the six mid-length stamens, which are graduated in length, have their anthers standing close above and close beneath the mouth of the calyx; the six shortest stamens rise rather above the middle of the calyx. In the mid-styled form the stigma projects just above the mouth of the calyx, and stands almost on a level with the mid-length stamens of the long- and short-styled forms; its own longest stamens project well above the mouth of the calyx, and stand a little above the level of the stigma of the long-styled form. In short, without entering on further details, there is a close general correspondence in structure between this species and *L. salicaria*, but with some differences in the proportional lengths of the parts. The fact of each of the three pistils having two sets of stamens of corresponding lengths, borne by the two other forms, comes out conspicuously. In the mid-styled form the pollen grains from the longest stamens are nearly double the diameter of those from the shortest stamens; so that there is a greater difference in this respect than in *L. salicaria*. In the long-styled form, also, the difference in diameter between the pollen grains of the mid-styled and shortest stamens is greater than in *L. salicaria*. These comparisons, however, must be received with caution, as they were made on specimens soaked in water, after having been long kept dry.

Lythrum thymifolia. This form, according to Vaucher,[8] is / dimorphic, like Primula, and therefore presents only two forms. I received two dried flowers from Kew, which consisted of the two forms; in one the stigma projected far beyond the calyx, in the other it was included within the calyx; in this latter form the style was only one-fourth of the length of that in the other form. There are only six stamens; these are somewhat graduated in length, and their anthers in the short-styled form stand a little above the stigma, but yet by no means equal in length the pistil of the long-styled form. In the latter the stamens are rather shorter than those in the other form. The six stamens alternate with the petals, and therefore correspond homologically with the longest stamens of *L. salicaria* and *L. Graefferi*.

Lythrum hyssopifolia. This species is said by Vaucher, but I believe erroneously, to be dimorphic. I have examined dried flowers from twenty-two separate plants from various localities, sent to me by Mr Hewett C. Watson, Professor Babington, and others. These were all essentially alike, so that the species

---

[8] *Hist. Phys. des Plantes d'Europe*, vol. ii (1841), pp. 369, 371.

cannot be heterostyled. The pistil varies somewhat in length, but when unusually long, the stamens are likewise generally long; in the bud the stamens are short; and Vaucher was perhaps thus deceived. There are from six to nine stamens graduated in length. The three stamens, which vary in being either present, or absent, correspond with the six shorter stamens of *L. salicaria* and with the six which are always absent in *L. thymefolia*. The stigma is included within the calyx, and stands in the midst of the anthers, and would generally be fertilized by them; but as the stigma and anthers are upturned, and as, according to Vaucher, there is a passage left in the upper side of the flower to the nectary, there can hardly be a doubt that the flowers are visited by insects, and would occasionally be cross-fertilized by them, as surely as the flowers of the short-styled *L. salicaria*, the pistil of which and the corresponding stamens in the other two forms closely resemble those of *L. hysopifolia*. According to Vaucher and Lecoq,[9] this species, which is an annual, generally grows almost solitarily, whereas the three preceding species are social; and this fact alone would almost have convinced me that *L. hyssopifolia* was not heterostyled, as such plants cannot habitually live isolated any better than one sex of a dioecious species. /

We thus see that within this genus some species are heterostyled and trimorphic; one apparently heterostyled and dimorphic, and one homostyled.

Nesaea verticillata. I raised a number of plants from seed sent me by Professor Asa Gray, and they presented three forms. These differed from one another in the proportional lengths of their organs of fructification and in all respects, in very nearly the same way as the three forms of *Lythrum Graefferi*. The green-pollen grains from the longest stamens, measured along their longer axis and not distended with water, were 13/7000 of an inch in length; those from the mid-length stamens 9–10/7000, and those from the shortest stamens 8–9/7000 of an inch. So that the largest pollen grains are to the smallest in diameter as 100 to 65. This plant inhabits swampy ground in the United States. According to Fritz Müller,[10] a species of this genus in St Catharina, in Southern Brazil, is homo-styled.

Lagerstroemia Indica. This plant, a member of the Lythraceae, may perhaps be heterostyled, or may formerly have been so. It is remarkable from the extreme variability of its stamens. On a plant, growing in my hothouse, the flowers included from nineteen to twenty-nine short stamens with yellow pollen, which correspond in position with the shortest stamens of Lythrum; and from one to five (the latter number being the commonest) very long stamens, with thick flesh-coloured filaments and green pollen, corresponding in position with the longest stamens of Lythrum. In one flower, two of the long stamens produced green, while a third produced yellow pollen, although the filaments of all three were thick and flesh-coloured. In an anther of another flower, one cell contained green and the other yellow pollen. The green and yellow pollen grains from the stamens of different length are of the same size. The pistil is a

[9] *Geograph. Bot. de l'Europe*, vol. vi, 1857, p. 157.
[10] *Bot. Zeitung*, 1868, p. 112.

little bowed upwards, with the stigma seated between the anthers of the short
and long stamens, so that this plant was mid-styled. Eight flowers were
fertilized with green pollen, and six with yellow pollen, but not one set fruit.
This latter fact by no means proves that the plant is heterostyled, as it may
belong to the class of self-sterile species. Another plant growing in the Botanic
Gardens at Calcutta, as Mr J. Scott informs me, was long-styled, and it was
equally / sterile with its own pollen; whilst a long-styled plant of *L. reginae*,
though growing by itself, produced fruit. I examined dried flowers from two
plants of *L. parviflora*, both of which were long-styled, and they differed from
L. Indica in having eight long stamens with thick filaments, and a crowd of
shorter stamens. Thus the evidence whether *L. Indica* is heterostyled is
curiously conflicting; the unequal number of the short and long stamens, their
extreme variability, and especially the fact of their pollen grains not differing
in size, are strongly opposed to this belief: on the other hand, the difference in
length of the pistils in two of the plants, their sterility with their own pollen,
and the difference in length and structure of the two sets of stamens in the
same flower, and in the colour of their pollen, favour the belief. We know that
when plants of any kind revert to a former condition, they are apt to be highly
variable, and the two halves of the same organ sometimes differ much, as in the
case of the above-described anther of the Lagerstroemia; we may therefore
suspect that this species was once heterostyled, and that it still retains traces of
its former state, together with a tendency to revert more completely to it. It
deserves notice, as bearing on the nature of Lagerstroemia, that in *Lythrum
hyssopifolia*, which is a homostyled species, some of the shorter stamens vary in
being either present or absent, and that these stamens are altogether absent in
L. thymifolia. In another genus of the Lythraceae, namely Cuphea, three species
raised by me from seed certainly were homostyled; nevertheless their stamens
consisted of two sets differing in length and in the colour and thickness of their
filaments, but not in the size or colour of their pollen grains; so that they thus
far resembled the stamens of Lagerstroemia. I found that *Cuphea purpurea* was
highly fertile with its own pollen when artificially aided, but sterile when insects
were excluded.[11] /

[11] Mr Spence informs me that in several species of the genus Mollia (Tiliaceae)
which he collected in South America, the stamens of the five outer cohorts have
purplish filaments and green pollen, whilst the stamens of the five inner cohorts
have yellow pollen. He therefore suspected that these species might proved to be
heterostyled and trimorphic: but he did not notice the length of the pistils. In the
allied Luhea the outer purplish stamens are destitute of anthers. I procured some
specimens of *Mollia lepidota* and *speciosa* from Kew, but could not make out that their
pistils differed in length in different plants; and in all those which I examined the
stigma stood close beneath the uppermost anthers. The numerous stamens are
graduated in length, and the pollen grains from the longest and shortest ones did
not present any marked difference in diameter. Therefore these species do not
appear to be heterostyled.

OXALIS (GERANIACEAE)

In 1863 Mr Roland Trimen wrote to me from the Cape of Good Hope that he had there found species of Oxalis which presented three forms; and of these he enclosed drawings and dried specimens. Of one species he collected 43 flowers from distinct plants, and they consisted of 10 long-styled, 12 mid-styled, and 21 short-styled. Of another species he collected 13 flowers, consisting of 3 long-styled, 7 mid-styled, and 3 short-styled. In 1866 Professor Hildebrand proved[12] by an examination of the specimens in several herbaria that 20 species are certainly heterostyled and trimorphic, and 51 others almost certainly so. He also made some interesting observations on living plants belonging to one form alone; for at that time he did not possess the three forms of any living species. During the years 1864 and 1868 I occasionally experimented on *Oxalis speciosa*, but until now have never found time to publish the results. In 1871 Hildebrand published an admirable paper[13] in which he shows in the case of two species of Oxalis, that the sexual relations of the three forms are nearly the same as in *Lythrum salicaria*. I will now give an abstract of his observations, and afterwards of my own less complete ones. I may premise that in all the species seen by me, the stigmas of the five straight pistils of the long-styled form stand on a level with the anthers of the longest

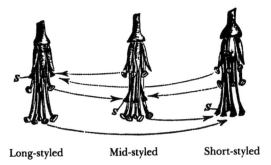

Long-styled   Mid-styled   Short-styled

Fig. 11   *Oxalis speciosa* (with the petals removed)

S S S, stigmas. The dotted lines with arrows show which pollen must be carried to the stigmas for legitimate fertilization.

[12] *Monatsber. der Akad. der Wiss. Berlin*, 1866, pp. 352, 372. He gives drawings of the three forms at p. 42 of his *Geschlechter-Vertheilung*, etc., 1867.

[13] *Bot. Zeitung*, 1871, pp. 416 and 432.

stamens in the two other forms. In the mid-styled / form, the stigmas pass out between the filaments of the longest stamens (as in the short-styled form of Linum); and they stand rather nearer to the upper anthers than to the lower ones. In the short-styled form, the stigmas also pass out between the filaments nearly on a level with the tips of the sepals. The anthers in this latter form and in the mid-styled rise to the same height as the corresponding stigmas in the other two forms.

Oxalis Valdiviana. This species, an inhabitant of the west coast of South America, bears yellow flowers. Hildebrand states that the stigmas of the three forms do not differ in any marked manner, but that the pistil of the short-styled form alone is destitute of hairs. The diameters of the pollen grains are as follows:

		Divisions of the micrometer
From the	longest stamens of short-styled	8 to 9
„	mid-length stamens of short-styled	7 to 8
„	longest stamens of mid-styled	8
„	shortest stamens of mid-styled	6
„	mid-length stamens of long-styled	7
„	shortest stamens of long-styled	6

Therefore the extreme difference in dimeter is as $8 \cdot 5$ to 6, or as 100 to 71. The results of Hildebrand's experiments are given in the following table, drawn up in accordance with my usual plan. He fertilized each form with pollen from the two sets of anthers of the same flower, and likewise from flowers on distinct plants belonging to the same form; but the effects of these two closely allied kinds of fertilization differ so little that I have not kept them distinct.

TABLE 27   *Oxalis Valdiviana (from Hildebrand)*

Nature of union	*Number of flowers fertilized*	*Number of capsules produced*	*Number of seeds per capsule*
Long-styled form, by pollen of longest stamens of short-styled. Legitimate union	28	28	11·9
Long-styled form, by pollen of longest stamens of mid-styled. Legitimate union	21	21	12·0
Long-styled form, by pollen of own and own-form mid-length stamens. Illegitimate union	40	2	5·5

TABLE 27   *Oxalis Valdiviana (from Hildebrand) – continued*

Nature of union	Number of flowers fertilized	Number of capsules produced	Number of seeds per capsule
Long-styled form, by pollen of own and own-form shortest stamens. Illegitimate union	26	0	0
Long-styled form, by pollen of shortest stamens of short-styled. Illegitimate union	16	1	1
Long-styled form, by pollen of shortest stamens. Illegitimate union	9	0	0 /
Mid-styled form, by pollen of mid-length stamens of long-styled. Legitimate union	38	38	11·3
Mid-styled form, by pollen of mid-length stamens of short-styled. Legitimate union	23	23	10·4
Mid-styled form, by pollen of own and own-form longest stamens. Illegitimate union	52	0	0
Mid-styled form, by pollen of own and own-form shortest stamens. Illegitimate union	30	1	6
Mid-styled form, by pollen of shortest stamens of long-styled. Illegitimate union	16	0	0
Mid-styled form, by pollen of longest stamens of short-styled. Illegitimate union	16	2	2·5
Short-styled form, by pollen of shortest stamens of long-styled. Legitimate union	18	18	11·0
Short-styled form, by pollen of shortest stamens of mid-styled. Legitimate union	10	10	11·3
Short-styled form, by pollen of own and own-form longest stamens. Illegitimate union	21	0	0
Short-styled form, by pollen of own and own-form mid-length stamens. Illegitimate union	22	0	0
Short-styled form, by pollen of longest stamens of mid-styled. Illegitimate union	4	0	0
Short-styled form, by pollen of mid-length stamens of long-styled. Illegitimate union	3	0	0 /

We here have the remarkable result that every one of 138 legitimately fertilized flowers on the three forms yielded capsules, containing on an average 11·33 seeds. Whilst of the 255 illegitimately fertilized flowers, only 6 yielded capsules, which contained 3·83 seeds on an average. Therefore the fertility of the six legitimate to that of the twelve illegitimate unions, as judged by the proportion of flowers that yielded capsules, is as 100 to 2, and as judged by the average number of seeds per capsule as 100 to 34. It may be added that some plants which were protected by nets did not spontaneously produce any fruit; nor did one which was left uncovered by itself and was visited by bees. On the other hand, scarcely a single flower on some uncovered plants of the three forms growing near together failed to produce fruit.

Oxalis Regnelli. This species bears white flowers and inhabits Southern Brazil. Hildebrand says that the stigma of the long-styled form is somewhat larger than that of the mid-styled, and this than that of the short-styled. The pistil of the latter is clothed with a few hairs, whilst it is very hairy in the other two forms. The diameter of the pollen grains from both sets of the longest stamens equals 9 divisions of the micrometer – that from the mid-length stamens of the long-styled form between 8 and 9, and of the short-styled 8 – and that from the shortest stamens of both sets 7. So that the extreme difference in diameter is as 9 to 7 or as 100 to 78. The experiments made by Hildebrand, which are not so numerous as in the last case, are given in Table 28 in the same manner as before.

TABLE 28   *Oxalis Regnelli (from Hildebrand)*

Nature of union	Number of flowers fertilized	Number of capsules produced	Average number of seeds per capsule
Long-styled form, by pollen of longest stamens of short-styled. Legitimate union	6	6	10·1
Long-styled form, by pollen of longest stamens of mid-styled. Legitimate union	5	5	10·6
Long-styled form, by pollen of own mid-length stamens. Illegitimate union	4	0	ɔ
Long-styled form, by pollen of own shortest stamen. Illegitimate union	1	0	0
Mid-styled form, by pollen of mid-length stamens of short-styled. Legitimate union	9	9	10·4
Mid-styled form, by pollen of mid-length stamens of long-styled. Legitimate union	10	10	10·1
Mid-styled form, by pollen of own longest stamens. Illegitimate union	9	0	0
Mid-styled form, by pollen of own shortest stamens. Illegitimate union	2	0	0
Mid-styled form, by pollen of longest stamens of short-styled. Illegitimate union	1	0	0

TABLE 28   *Oxalis Regnelli (from Hildebrand) – continued*

Nature of union	Number of flowers fertilized	Number of capsules produced	Average number of seeds per capsule
Short-styled form, by pollen of shortest stamens of mid-styled. Legitimate union	9	9	10·6
Short-styled form, by pollen of shortest stamens of long-styled. Legitimate union	2	2	9·5
Short-styled form, by pollen of own mid-length stamens. Illegitimate union	12	0	0
Short-styled form, by pollen of own longest stamens. Illegitimate union	9	0	0
Short-styled form, by pollen of mid-length stamens of long-styled. Illegitimate union	1	0	0

The results are nearly the same as in the last case, but more striking; for 41 flowers belonging to the three forms fertilized legitimately all yielded capsules, / containing on an average 10·31 seeds; whilst 39 flowers fertilized illegitimately did not yield a single capsule or seed. Therefore the fertility of the six legitimate to that of the several illegitimate unions, as judged both by the proportion of flowers which yielded capsules and by the average number of contained seeds, is as 100 to 0.

Oxalis speciosa. This species, which bears pink flowers, was introduced from the Cape of Good Hope. A sketch of the reproductive organs of the three forms (Fig. 11) has already been given. The stigma of the long-styled form (with the papillae on its surface included) is twice as large as that of the short-styled, and that of the mid-styled intermediate in size. The pollen grains from the stamens in the three forms are in their longer diameters as follows:

	Divisions of the micrometer
From the longest stamens of short-styled	15 to 16
„ mid-length stamens of short-styled	12 „ 13
„ longest stamens of mid-styled	16
„ shortest stamens of mid-styled	11 to 12
„ mid-length stamens of long-styled	14
„ shortest stamens of long-styled	12

Therefore the extreme difference in diameter is as 16 to 11, or as 100 to 69; but as the measurements were taken at different times, they are probably only approximately accurate. The results of my experiments in fertilizing the three forms are given in the following table. /

TABLE 29   *Oxalis speciosa*

Nature of union	Number of flowers fertilized	Number of capsules produced	Average number of seeds per capsule
Long-styled form, by pollen of longest stamens of short-styled. Legitimate union	19	15	57·4 '
Long-styled form, by pollen of longest stamens of mid-styled. Legitimate union	4	3	59·0
Long-styled form, by pollen of own-form mid-length stamens. Illegitimate union	9	2	42·5
Long-styled form, by pollen of own-form shortest stamens. Illegitimate union	11	0	0
Long-styled form, by pollen of shortest stamens of mid-styled. Illegitimate union	4	0	0
Long-styled form by pollen of mid-length stamens of short-styled. Illegitimate union	12	5	30·0
Mid-styled form, by pollen of mid-length stamens of long-styled. Legitimate union	3	3	63·6
Mid-styled form, by pollen of mid-length stamens of short-styled. Legitimate union	4	4	56·3
Mid-styled form, by mixed pollen from both own-form longest and shortest stamens. Illegitimate union	9	2	19
Mid-styled form, by pollen of longest stamens of short-styled. Illegitimate union	12	1	8 /
Short-styled form, by pollen of shortest stamens of mid-styled. Legitimate union	3	2	67
Short-styled form, by pollen of shortest stamens of long-styled. Legitimate union	3	3	54·3
Short-styled form, by pollen of own-form longest stamens. Illegitimate union	5	1	8
Short-styled form, by pollen of own-form mid-length stamens. Illegitimate union	3	0	0
Short-styled form, by both pollens mixed together, of own-form longest and mid-length stamens. Illegitimate union	13	0	0
Short-styled form, by pollen of longest stamens of mid-styled. Illegitimate union	7	0	0
Short-styled form, by pollen of mid-length stamens of long-styled. Illegitimate union	10	1	54

We here see that thirty-six flowers on the three forms legitimately fertilized yielded 30 capsules, these containing on an average 58·36 seeds. Ninety-five flowers illegitimately fertilized yielded 12 capsules, containing on an average 28·58 seeds. Therefore the fertility of the six legitimate to that of the twelve illegitimate unions, as judged by the proportion of flowers which yielded capsules, is as 100 to 15, and judged by the average number of seeds per capsule as 100 to 49. This plant, in comparison with the two South American species previously described, produces / many more seeds, and the illegitimately fertilized flowers are not quite so sterile.

Oxalis rosea. Hildebrand possessed in a living state only the long-styled form of this trimorphic Chilian species.[14] The pollen grains from the two sets of anthers differ in diameter as 9 to 7·5, or as 100 to 83. He has further shown that there is an analogous difference between the grains from the two sets of anthers of the same flower in five species of Oxalis, besides those already decribed. The present species differs remarkably from the long-styled form of the three species previously experimented on, in a much larger proportion of the flowers setting capsules when fertilized with their own-form pollen. Hildebrand fertilized 60 flowers with pollen from the mid-length stamens (of either the same or another flower), and they yielded no less than 55 capsules, or 92 per cent. These capsules contained on an average 5·62 seeds; but we have no means of judging how near an approach this average makes to that from flowers legitimately fertilized. He also fertilized 45 flowers with pollen from the shortest stamens, and these yielded only 17 capsules, or 31 per cent, containing on an average only 2·65 seeds. We thus see that about thrice as many flowers, when fertilized with pollen from the mid-length stamens, produced capsules, and these contained twice as many seeds, as did the flowers fertilized with pollen from the shortest stamens. It thus appears (and we find some evidence of the same fact with *O. speciosa*), that the same rule holds good with Oxalis as with *Lythrum salicaria*; namely, that in any two unions, the greater the inequality in length between the pistils and stamens, or, / which is the same thing, the greater the distance of the stigma from the anthers, the pollen of which is used for fertilization, the less fertile is the union – whether judged by the proportion of flowers which set capsules, or by the average number of seeds per

[14] *Monatsber. der Akad. der Wiss. Berlin*, 1866, p. 372.

capsule. The rule cannot be explained in this case any more than in that of Lythrum, by supposing that wherever there is greater liability to self-fertilization, this is checked by the union being rendered more sterile; for exactly the reverse occurs, the liability to self-fertilization being greatest in the unions between the pistils and stamens which approach each other the nearest, and these are the more fertile. I may add that I also possessed some long-styled plants of this species: one was covered by a net, and it set spontaneously a few capsules, though extremely few compared with those produced by a plant growing by itself, but exposed to the visits of bees.

With most of the species of Oxalis the short-styled form seems to be the most sterile of the three forms, when these are illegitimately fertilized; and I will add two other cases to those already given. I fertilized 29 short-styled flowers of *O. compressa* with pollen from their own two sets of stamens (the pollen grains of which differ in diameter. as 100 and 83), and not one produced a capsule. I formerly cultivated during several years the short-styled form of a species purchased under the name of *O. Duwii* (but I have some doubts whether it was rightly named), and fertilized many flowers with their own two kinds of pollen, which differ in diameter in the usual manner, but never got a single seed. On the other hand, Hildebrand says that the short-styled form of *O. Deppei*, growing by itself, yields plenty of seed; but it is not positively known that this species is heterostyled; and / the pollen grains from the two sets of anthers do not differ in diameter.

Some facts communicated to me by Fritz Müller afford excellent evidence of the utter sterility of one of the forms of certain trimorphic species of Oxalis, when growing isolated. He has seen in St Catharina, in Brazil, a large field of young sugar cane, many acres in extent, covered with the red blossoms of one form alone, and these did not produce a single seed. His own land is covered with the short-styled form of a white-flowered trimorphic species, and this is equally sterile; but when the three forms were planted near together in his garden they seeded freely. With two other trimorphic species he finds that isolated plants are always sterile.

Fritz Müller formerly believed that a species of Oxalis, which is so abundant in St Catharina that it borders the roads for miles, was dimorphic instead of trimorphic. Although the pistils and stamens vary greatly in length, as was evident in some specimens sent to me, yet the plants can be divided into two sets, according to the lengths of these organs. A large proportion of the anthers are of a white colour and quite destitute of pollen; others which are pale yellow contain many bad with some good grains; and others again which are bright yellow have apparently sound pollen; but he has never succeeded in finding any fruit on this species. The stamens in some of the flowers are partially converted into petals. Fritz Müller after reading my description, hereafter to be given, of the illegitimate offspring of various heterostyled species, suspects that these plants of Oxalis may be the variable and sterile offspring of a single form of some trimorphic species, perhaps accidentally introduced into the district, which has since been / propagated asexually. It is probable that this kind of propagation would be much aided by there being no expenditure in the production of seed.

Oxalis (Biophytum) sensitiva. This plant is ranked by many botanists as a distinct genus. Mr Thwaites sent me a number of flowers preserved in spirits from Ceylon, and they are clearly trimorphic. The style of the long-styled form is clothed with many scattered hairs, both simple and glandular; such hairs are much fewer on the style of the mid-styled, and quite absent from that of the short-styled form; so that this plant resembles in this respect *O. Valdiviana* and *Regnelli.* Calling the length of the two lobes of the stigma of the long-styled form 100, that of the mid-styled is 141, and that of the short-styled 164. In all other cases, in which the stigma in this genus differs in size in the three forms, the difference is of a reversed nature, the stigma of the long-styled being the largest, and that of the short-styled the smallest. The diameter of the pollen grains from the longest stamens being represented by 100, those from the mid-length stamens are 91, and those from the shortest stamens 84 in diameter. This plant is remarkable, as we shall see in the last chapter of this volume, by producing long-styled, mid-styled, and short-styled cleistogamic flowers.

Homostyled species of Oxalis

Although the majority of the species in the large genus Oxalis seem to be trimorphic, some are homostyled, that is, exist under a single form; for instance, the common *O. acetosella,* and according to Hildebrand two other widely distributed European species, *O. stricta* and *corniculata.* Fritz Müller also informs me that a similarly constituted species is found in St Catharina, and that it is / quite fertile with its own pollen when insects are excluded. The stigmas of *O. stricta* and of another homostyled species, viz. *O. tropaeoloides*, commonly stand on a level with the upper anthers, and both these species are likewise quite fertile when insects are excluded.

With respect to *O. acetosella*, Hildebrand says that in all the many specimens examined by him the pistil exceeded the longer stamens in length. I procured 108 flowers from the same number of plants growing in three distant parts of England; of these 86 had their stigmas projecting considerably above, whilst 22 had them nearly on a level with the upper anthers. In one lot of 17 flowers from the same wood, the stigmas in every flower projected fully as much above the upper anthers as these stood above the lower anthers. So that these plants might fairly be compared with the long-styled form of a heterostyled species; and I at first thought that *O. acetosella* was trimorphic. But the case is one merely of great variability. The pollen grains from the two sets of anthers, as observed by Hildebrand and myself, do not differ in diameter. I fertilized twelve flowers on several plants with pollen from a distinct plant, choosing those with pistils of a different length; and 10 of these (i.e. 83 per cent) produced capsules, which contained on an average 7·9 seeds. Fourteen flowers were fertilized with their own pollen, and 11 of these (i.e. 79 per cent) yielded capsules, containing a larger average of seed, namely 9·2. These plants, therefore, in function show not the least sign of being heterostyled. I may add that 18 flowers protected by a net were left to fertilize themselves, and only 10 of these (i.e. 55 per cent) yielded capsules, which contained on an average only 6·3 seeds. So that the access of insects, or artificial aid in placing pollen on the stigma, increases the fertility of the / flowers; and I found that this applied especially to those having shorter pistils. It should be remembered that the flowers hang downwards, so that those with short pistils would be the least likely to receive their own pollen, unless they were aided in some manner.

Finally, as Hildebrand has remarked, there is no evidence that any of the heterostyled species of Oxalis are tending towards a dioecious condition, as Zuccarini and Lindley inferred from the differences in the reproductive organs of the three forms, the meaning of which they did not understand.

PONTEDERIA [SP.?] (PONTEDERIACEAE)

Fritz Müller found this aquatic plant, which is allied to the Liliaceae,

131

growing in the greatest profusion on the banks of a river in Southern Brazil.[15] But only two forms were found, the flowers of which include three long and three short stamens. The pistil of the long-styled form, in two dried flowers which were sent me, was in length as 100 to 32, and its stigma as 100 to 80, compared with the same organs in the short-styled form. The long-styled stigma projects considerably above the upper anthers of the same flower, and stands on a level with the upper ones of the short-styled form. In the latter the stigma is seated beneath both its own sets of anthers, and is on a level with the anthers of the shorter stamens in the long-styled form. The anthers of the longer stamens of the short-styled form are to those of the shorter stamens of the long-styled form as 100 to 88 in length. The pollen grains distended / with water from the longer stamens of the short-styled form are to those from the shorter stamens of the same form as 100 to 87 in diameter, as deduced from ten measurements of each kind. We thus see that the organs in these two forms differ from one another and are arranged in an analogous manner, as in the long- and short-styled forms of the trimorphic species of Lythrum and Oxalis. Moreover, the longer stamens of the long-styled form of Pontederia, and the shorter ones of the short-styled form are placed in a proper position for fertilizing the stigma of a mid-styled form. But Fritz Müller, although he examined a vast number of plants, could never find one belonging to the mid-styled form. The older flowers of the long-styled and short-styled plants had set plenty of apparently good fruit; and this might have been expected, as they could legitimately fertilize one another. Although he could not find the mid-styled form of this species, he possessed plants of another species growing in his garden, and all these were mid-styled; and in this case the pollen grains from the anthers of the longer stamens were to those from the shorter stamens of the same flower as 100 to 86 in diameter, as deduced from ten measurements of each kind. These mid-styled plants growing by themselves never produced a single fruit.

Considering these several facts, there can hardly be a doubt that both these species of Pontederia are heterostyled and trimorphic. This cause is an interesting one, for no other monocotyledonous plant is known to be heterostyled. Moreover, the flowers are irregular, and all other heterostyled plants have almost symmetrical flowers. The two

[15] 'Ueber den Trimorphismus der Pontederien'; *Jenaische Zeitschrift*, etc., vol. 6, 1871, p. 74.

forms differ somewhat in the colour of their corollas, that of the short-styled being of a darker blue, whilst that of the long-styled / tends towards violet, and no other such case is known. Lastly, the three longer stamens alternate with the three shorter ones, whereas in Lythrum and Oxalis the long and short stamens belong to distinct whorls. With respect to the absence of the mid-styled form in the case of the Pontederia which grows wild in Southern Brazil, this would probably follow if only two forms had been originally introduced there; for, as we shall hereafter see from the observations of Hildebrand, Fritz Müller and myself, when one form of Oxalis is fertilized exclusively by either of the other two forms, the offspring generally belong to the two parent forms.

Fritz Müller has recently discovered, as he informs me, a third species of Pontederia, with all three forms growing together in pools in the interior of S. Brazil; so that no shadow of doubt can any longer remain about this genus including trimorphic species. He sent me dried flowers of all three forms. In the long-styled form the stigma stands a little above the tips of the petals, and on a level with the anthers of the longest stamens in the other two forms. The pistil is in length to that of the mid-styled as 100 to 56, and to that of the short-styled as 100 to 16. Its summit is rectangularly bent upwards, and the stigma is rather broader than that of the mid-styled, and broader in about the ratio of 7 to 4 than that of the short-styled. In the mid-styled form, the stigma is placed rather above the middle of the corolla, and nearly on a level with the mid-length stamens in the other two forms: its summit is a little bent upwards. In the short-styled form the pistil is, as we have seen, very short, and differs from that in the other two forms in being straight. It stands rather beneath the level of the anthers of the shortest stamens in the long-styled and / mid-styled forms. The three anthers of each set of stamens, more especially those of the shortest stamens, are placed one beneath the other, and the ends of the filaments are bowed a little upwards, so that the pollen from all the anthers would be effectively brushed off by the proboscis of a visiting insect. The relative diameters of the pollen grains, after having been long soaked in water, are given in the following list (p. 134), as measured by my son Francis.

We have here the usual rule of the grains from the longer stamens, the tubes of which have to penetrate the longer pistil, being larger than those from the stamens of less length. The extreme difference in diameter between the grains from the longest stamens of the mid-

	Divisions of the micrometer
Long-styled form, from the mid-length stamens	13·2
(Average of 20 measurements)	
Long-styled form, from the shortest stamens	9·0
(10 measurements)	
Mid-styled form, from the longest stamens	16·4
(15 measurements)	
Mid-styled form, from the shortest stamens	9·1
(20 measurements)	
Short-styled form, from the longest stamens	14·6
(20 measurements)	
Short-styled form, from the mid-length stamens	12·3
(20 measurements)	

styled form, and from the shortest stamens of the long-styled, is as 16·4 to 9·0, or as 100 to 55; and this is the greatest difference observed by me in any heterostyled plant. It is a singular fact that the grains from the corresponding longest stamens in the two forms differ considerably in diameter; as do those in a lesser degree from the corresponding mid-length stamens in the two forms; whilst those from the corresponding shortest stamens in the long- and mid-styled forms are almost exactly equal. Their inequality in the two first cases depends on the grains / in both sets of anthers in the short-styled form being smaller than those from the corresponding anthers in the other two forms; and here we have a case parallel with that of the mid-styled form of *Lythrum salicaria*. In this latter plant the pollen grains of the mid-styled forms are of smaller size and have less fertilizing power than the corresponding ones in the other two forms; whilst the ovarium, however fertilized, yields a greater number of seeds; so that the mid-styled form is altogether more feminine in nature than the other two forms. In the case of Pontederia, the ovarium includes only a single ovule, and what the meaning of the difference in size between the pollen grains from the corresponding sets of anthers may be, I will not pretend to conjecture.

The clear evidence that the species just described is heterostyled and trimorphic is the more valuable as there is some doubt with respect to *P. cordata*, an inhabitant of the United States. Mr Leggett suspects[16] that it is either dimorphic or trimorphic, for the pollen grains of the longer stamens are 'more than twice the diameter or than eight times

[16] *Bull. of the Torrey Botanical Club*, 1875, vol. vi, p. 62.

the mass of the grains of the shorter stamens. Though minute, these smaller grains seem as perfect as the larger ones'. On the other hand, he says that in all the mature flowers, 'the style was as long at least as the longer stamens'; 'whilst in the young flowers it was intermediate in length between the two sets of stamens'; and if this be so, the species can hardly be heterostyled. /

CHAPTER V

ILLEGITIMATE OFFSPRING OF
HETEROSTYLED PLANTS

Illegitimate offspring from all three forms of *Lythrum salicaria* – Their dwarfed stature and sterility, some utterly barren, some fertile – Oxalis, transmission of form to the legitimate and illegitimate seedlings – *Primula Sinensis*, illegitimate offspring in some degree dwarfed and infertile – Equal-styled varieties of *P. Sinensis, auricula, farinosa*, and *elatior* – *P. vulgaris*, red-flowered variety, illegitimate seedlings sterile – *P. veris*, illegitimate plants raised during several successive generations, their dwarfed stature and sterility – Equal-styled varieties of *P. veris* – Transmission of form by Pulmonaria and Polygonum – Concluding remarks – Close parallelism between legitimate fertilization and hybridism.

We have hitherto treated of the fertility of the flowers of heterostyled plants, when legitimately and illegitimately fertilized. The present chapter will be devoted to the character of their offspring or seedlings. Those raised from legitimately fertilized seeds will be here called *legitimate seedlings* or *plants*, and those from illegitimately fertilised seeds, *illegitimate seedlings* or *plants*. They differ chiefly in their degree of fertility, and in their powers of growth or vigour. I will begin with trimorphic plants, and I must remind the reader that each of the three forms can be fertilized in six different ways; so that all three together can be fertilized in eighteen different ways. For instance, a long-styled form can be fertilized legitimately by the longest stamens of the mid-styled and short-styled forms, and illegitimately by its own-form mid-length and shortest stamens, also by the mid-length stamens of the mid-styled and by the shortest stamens of the / short-styled form; so that the long-styled can be fertilized legitimately in two ways and illegitimately in four ways. The same holds good with respect to the mid-styled and short-styled forms. Therefore with trimorphic species six of the eighteen unions yield legitimate offspring, and twelve yield illegitimate offspring.

I will give the results of my experiments in detail, partly because the observations are extremely troublesome, and will not probably soon be repeated – thus, I was compelled to count under the microscope above 20,000 seeds of *Lythrum salicaria* – but chiefly because light is thus indirectly thrown on the important subject of hybridism.

LYTHRUM SALICARIA

Of the twelve illegitimate unions two were completely barren, so that no seeds were obtained, and of course no seedlings could be raised. Seedlings were, however, raised from seven of the ten remaining illegitimate unions. Such illegitimate seedlings when in flower were generally allowed to be freely and legitimately fertilized, through the agency of bees, by other illegitimate plants belonging to the two other forms growing close by. This is the fairest plan, and was usually followed; but in several cases (which will always be stated) illegitimate plants were fertilized with pollen taken from legitimate plants belonging to the other two forms; and this, as might have been expected, increased their fertility. *Lythrum salicaria* is much affected in its fertility by the nature of the season; and to avoid error from this source, as far as possible, my observations were continued during several years. Some few experiments were / tried in 1863. The summer of 1864 was too hot and dry, and though the plants were copiously watered, some few apparently suffered in their fertility, whilst others were not in the least affected. The years 1865 and, especially, 1866, were highly favourable. Only a few observations were made during 1867. The results are arranged in classes according to the parentage of the plants. In each case the average number of seeds per capsule is given, generally taken from ten capsules, which, according to my experience, is a nearly sufficient number. The maximum number of seeds in any one capsule is also given; and this is a useful point of comparison with the normal standard – that is, with the number of seeds produced by legitimate plants legitimately fertilized. I will give likewise in each case the minimum number. When the maximum and minimum differ greatly, if no remark is made on the subject, it may be understood that the extremes are so closely connected by intermediate figures that the average is a fair one. Large capsules were always selected for counting, in order to avoid over-estimating the infertility of the several illegitimate plants.

In order to judge of the degree of inferiority in fertility of the several illegitimate plants, the following statement of the average and of the maximum number of seeds produced by ordinary or legitimate plants, when legitimately fertilized, some artificially and some naturally, will serve as a standard of comparison, and may in each case be referred to. But I give under each experiment the percentage of seeds produced by the illegitimate plants, in comparison with the standard legitimate number of the same form. For instance, ten capsules from the illegitimate long-styled plant (No. 10), which was legitimately / and naturally fertilized by other illegitimate plants, contained on an average 44·2 seeds; whereas the capsules on legitimate long-styled plants, legitimately and naturally fertilized by other legitimate plants, contained on an average 93 seeds. Therefore this illegitimate plant yielded only 47 per cent of the full and normal complement of seeds.

Standard number of seeds produced by legitimate plants of the three forms, when legitimately fertilized

Long-styled form: average number of seeds in each capsule	93
maximum number observed out of twenty-three capsules	159
Mid-styled form: average number of seeds	130
maximum number observed out of thirty-one capsules	151
Short-styled form: average number of seeds	83·5
but we may, for the sake of brevity	say 83
maximum number observed out of twenty-five capsules	112

CLASSES I AND II

Illegitimate plants raised from long-styled parents fertilized with pollen from the mid-length or the shortest stamens of other plants of the same form

From this union I raised at different times three lots of illegitimate seedlings, amounting altogether to 56 plants. I must premise that, from not foreseeing the result, I did not keep a memorandum whether the eight plants of the first lot were the product of the mid-length or shortest stamens of the same form; but I have good reason to believe that they were the product of the latter. These eight plants were much more dwarfed, and much more sterile than those in the other two lots. The latter were raised from a long-styled / plant growing quite isolated, and fertilized by the agency of bees with its own pollen: and it is almost certain, from the relative position of the organs of fructification, that the stigma under these circumstances would receive pollen from the mid-length stamens.

All the fifty-six plants in these three lots proved long-styled; now, if the parent plants had been legitimately fertilized by pollen from the longest stamens of the mid-styled and short-styled forms, only about one-third of the seedlings would have been long-styled, the other two-thirds being mid-styled and short-styled. In some other trimorphic and dimorphic genera we shall find the same curious fact, namely, that the long-styled form, fertilized illegitimately by its own-form pollen, produces almost exclusively long-styled seedlings.[1]

The eight plants of the first lot were of low stature: three which I measured attained, when fully grown, the heights of only 28, 29, and 47 inches; whilst legitimate plants growing close by were double this height, one being 77 inches. They all betrayed in their general appearance a weak constitution; they flowered rather later in the season, and at a later age than ordinary plants. Some did not flower every year; and one plant, behaving in an unprecedented manner, did not flower until three years old. In the two other lots none of the plants grew quite to their full and proper height, as could at once be seen by comparing them with the adjoining rows of legitimate plants. In several plants in all three lots, many of the anthers were either shrivelled or contained brown and tough, or pulpy / matter, without any good pollen grains, and they never shed their contents; they were in the state designated by Gärtner[2] as contabescent, which term I will for the future use. In one flower all the anthers were contabescent excepting two which appeared to the naked eye sound; but under the microscope about two-thirds of the pollen grains were seen to be small and shrivelled. In another plant, in which all the anthers appeared sound, many of the pollen grains were shrivelled and of unequal sizes. I counted the seeds produced by seven plants (1 to 7) in the first lot of eight plants, probably the product of parents fertilized by their own-form shortest stamens, and the seeds produced by three plants in the other two lots, almost certainly the product of parents fertilized by their own-form mid-length stamens.

Plant 1. This long-styled plant was allowed during 1863 to be freely and legitimately fertilized by an adjoining illegitimate mid-styled plant, but it did

---

[1] Hildebrand first called attention (*Bot. Zeitung*, 1 January, 1864, p. 5) to this fact in the case of *Primula Sinensis*; but his results were not nearly so uniform as mine.

[2] *Beiträge zur Kenntniss der Befruchtung*, 1844, p. 116.

not yield a single seed-capsule. It was then removed and planted in a remote place close to a brother long-styled plant No. 2, so that it must have been freely though illegitimately fertilized; under these circumstances it did not yield during 1864 and 1865 a single capsule. I should here state that a legitimate or ordinary long-styled plant, when growing isolated, and freely though illegitimately fertilized by insects with its own pollen, yielded an immense number of capsules, which contained on an average 21·5 seeds.

Plant 2. This long-styled plant, after flowering during 1863 close to an illegitimate mid-styled plant, produced less than twenty capsules, which contained on an average between four and five seeds. When subsequently growing in company with No. 1, by which it will have been illegitimately fertilized, it yielded in 1866 not a single capsule, but in 1865 it yielded twenty-two capsules: the best of these, fifteen in number, were examined; eight contained no seed, and the remaining seven contained on an average only three seeds, and these seeds were / so small and shrivelled that I doubt whether they would have germinated.

Plants 3 and 4. These two long-styled plants, after being freely and legitimately fertilized during 1863 by the same illegitimate mid-styled plant as in the last case, were as miserably sterile as No. 2.

Plant 5. This long-styled plant, after flowering in 1863 close to an illegitimate mid-styled plant, yielded only four capsules, which altogether included only five seeds. During 1864, 1865, and 1866, it was surrounded either by illegitimate or legitimate plants of the other two forms; but it did not yield a single capsule. It was a superfluous experiment, but I likewise artificially fertilized in a legitimate manner twelve flowers; but not one of these produced a capsule; so that this plant was almost absolutely barren.

Plant 6. This long-styled plant, after flowering during the favourable year of 1866, surrounded by illegitimate plants of the other two forms, did not produce a single capsule.

Plant 7. This long-styled plant was the most fertile of the eight plants of the first lot. During 1865 it was surrounded by illegitimate plants of various parentage, many of which were highly fertile, and must thus have been legitimately fertilized. It produced a good many capsules, ten of which yielded an average of 36·1 seeds, with a maximum of 47 and a minimum of 22; so that this plant produced 39 per cent of the full number of seeds. During 1864 it was surrounded by legitimate and illegitimate plants of the other two forms; and nine capsules (one poor one being rejected) yielded an average of 41·9 seeds, with a maximum of 56 and a minimum of 28; so that, under these favourable circumstances, this plant, the most fertile of the first lot, did not yield, when legitimately fertilized, quite 45 per cent of the full complement of seeds.

In the second lot of plants in the present class descended from the long-styled form, almost certainly fertilized with pollen from its own mid-length stamens, the plants, as already stated, were not nearly so dwarfed or so sterile as in the first lot. All produced plenty of capsules. I counted the number of seeds in only three plants, viz. Nos 8, 9, and 10. /

Plant 8. This plant was allowed to be freely fertilized in 1864 by legitimate and illegitimate plants of the other two forms, and ten capsules yielded on an average 41·1 seeds, with a maximum of 73 and a minimum of 11. Hence this plant produced only 44 per cent of the full complement of seeds.

Plant 9. This long-styled plant was allowed in 1865 to be freely fertilized by illegitimate plants of the other two forms, most of which were moderately fertile. Fifteen capsules yielded on an average 57·1 seeds, with a maximum of 86 and a minimum of 23. Hence the plant yielded 68 per cent of the full complement of seeds.

Plant 10. This long-styled plant was freely fertilized at the same time and in the same manner as the last. Ten capsules yielded an average of 44·2 seeds, with a maximum of 69 and a minimum of 25; hence this plant yielded 47 per cent of the full complement of seeds.

The nineteen long-styled plants of the third lot, of the same parentage as the last lot, were treated differently; for they flowered during 1867 by themselves so that they must have been illegitimately fertilized by one another. It has already been stated that a legitimate long-styled plant, growing by itself and visited by insects, yielded an average of 21·5 seeds per capsule, with a maximum of 35; but, to judge fairly of its fertility, it ought to have been observed during successive seasons. We may also infer from analogy that, if several legitimate long-styled plants were to fertilize one another, the average number of seeds would be increased; but how much increased I do not know; hence I have no perfectly fair standard of comparison by which to judge of the fertility of the three following plants of the present lot, the seeds of which I counted.

Plant 11. This long-styled plant produced a large crop of capsules, and in this respect was one of the most fertile of the whole lot of nineteen plants. But the average from ten / capsules was only 35·9 seeds, with a maximum of 60 and a minimum of 8.

Plant 12. This long-styled plant produced very few capsules, and ten yielded an average of only 15·4 seeds, with a maximum of 30 and a minimum of 4.

Plant 13. This plant offers an anomalous case; it flowered profusely, yet produced very few capsules; but these contained numerous seeds. Ten capsules yielded an average of 71·9 seeds, with a maximum of 95 and a minimum of 29. Considering that this plant was illegitimate and illegitimately fertilized by its brother long-styled seedlings, the average and the maximum are so remarkably high that I cannot at all understand the case. We should remember that the average for a legitimate plant legitimately fertilized is 93 seeds.

CLASS III

Illegitimate plants raised from a short-styled parent fertilized with pollen from own-form mid-length stamens

I raised from this union nine plants, of which eight were short-styled and one long-styled; so that there seems to be a strong tendency in this form to reproduce, when self-fertilized, the parent form; but the tendency is not so strong as with the long-styled. These nine plants never attained the full height of legitimate plants growing close to them. The anthers were contrabescent in many of the flowers on several plants.

Plant 14. This short-styled plant was allowed during 1865 to be freely and legitimately fertilized by illegitimate plants descended from self-fertilized mid-, long-, and short-styled plants. Fifteen capsules yielded an average of 28·3 seeds, with a maximum of 51 and a minimum of 11; hence this plant produced only 33 per cent of the proper number of seeds. The seeds themselves were small and irregular in shape. Although so sterile on the female side, none of the anthers were contabescent.

Plant 15. This short-styled plant, treated like the last during / the same year, yielded an average, from fifteen capsules, of 27 seeds, with a maximum of 49 and a minimum of 7. But two poor capsules may be rejected, and then the average rises to 32·6, with the same maximum of 49 and a minimum of 20; so that this plant attained 38 per cent of the normal standard of fertility, and was rather more fertile than the last, yet many of the anthers were contabescent.

Plant 16. This short-styled plant, treated like the two last, yielded from ten capsules an average of 77·8 seeds, with a maximum of 97 and a minimum of 60; so that this plant produced 94 per cent of the full number of seeds.

Plant 17. This, the one long-styled plant of the same parentage as the last three plants, when freely and legitimately fertilized in the same manner as the last, yielded an average from ten capsules of 76·3 rather poor seeds, with a maximum of 88 and a minimum of 57. Hence this plant produced 82 per cent of the proper number of seeds. Twelve flowers enclosed in a net were artificially and legitimately fertilized with pollen from a legitimate short-styled plant; and nine capsules yielded an average of 82·5 seeds, with a maximum of 98 and a minimum of 51; so that its fertility was increased by the action of pollen from a legitimate plant, but still did not reach the normal standard.

CLASS IV

Illegitimate plants raised from a mid-styled parent fertilized with pollen from own-form longest stamens

After two trials, I succeeded in raising only four plants from this illegitimate union. These proved to be three mid-styled and one long-styled; but from so small a number we can hardly judge of the tendency in mid-styled plants when self-fertilized to reproduce the same form. These four plants never attained their full and normal height; the long-styled plant had several of its anthers contabescent.

Plant 18. This mid-styled plant, when freely and legitimately fertilized during 1865 by illegitimate plants descended from self-fertilized long-, short-, and mid-styled plants, yielded an average from ten capsules of 102·6 seeds, with a maximum of / 131 and a minimum of 63: hence this plant did not produce quite 80 per cent of the normal number of seeds. Twelve flowers were artificially and legitimately fertilized with pollen from a legitimate long-styled plant, and yielded from nine capsules an average of 116·1 seeds, which were finer than in the previous case, with a maximum of 135 and a minimum of 75; so that, as with Plant 17, pollen from a legitimate plant increased the fertility, but did not bring it up to the full standard.

Plant 19. This mid-styled plant, fertilized in the same manner and at the same period as the last, yielded an average from ten capsules of 73·4 seeds, with a maximum of 87 and a minimum of 64: hence this plant produced only 56 per cent of the full number of seeds. Thirteen flowers were artificially and legitimately fertilized with pollen from a legitimate long-styled plant, and yielded ten capsules with an average of 95·6 seeds; so that the application of pollen from a legitimate plant added, as in the two previous cases, to the fertility, but did not bring it up to the proper standard.

Plant 20. This long-styled plant, of the same parentage with the two last mid-styled plants, and freely fertilized in the same manner, yielded an average from ten capsules of 69·6 seeds, with a maximum of 83 and a minimum of 52: hence this plant produced 75 per cent of the full number of seeds.

CLASS V

Illegitimate plants raised from a short-styled parent fertilized with pollen from the mid-length stamens of the long-styled form

In the four previous classes, plants raised from the three forms fertilized with pollen from either the longer or shorter stamens of the same form, but generally not from the same plant, have been described. Six other illegitimate unions are possible, namely, between the three forms and the stamens in the other two forms which do not correspond in height, with their pistils. But I succeeded in raising plants from only three of these six unions. From one of them, forming the present Class V, twelve plants were raised; these consisted of eight short-styled, and four long-styled plants, / with not one mid-styled.

These twelve plants never attained quite their full and proper height, but by no means deserved to be called dwarfs. The anthers in some of the flowers were contabescent. One plant was remarkable from all the longer stamens in every flower and from many of the shorter ones having their anthers in this condition. The pollen of four other plants, in which none of the anthers were contabescent, was examined; in one a moderate number of grains were minute and shrivelled, but in the other three they appeared perfectly sound. With respect to the power of producing seed, five plants (Nos 21 to 25) were observed: one yielded scarcely more than half the normal number; a second was slightly infertile; but the three others actually produced a larger average number of seeds, with a higher maximum, than the standard. In my concluding remarks I shall recur to this fact, which at first appears inexplicable.

Plant 21. This short-styled plant, freely and legitimately fertilized during 1865 by illegitimate plants, descended from self-fertilized long-, mid,- and short-styled parents, yielded an average from ten capsules of 43 seeds, with a maximum of 63 and a minimum of 26: hence this plant, which was the one with all its longer and many of its shorter stamens contabescent, produced only 52 per cent of the proper number of seeds.

Plant 22. This short-styled plant produced perfectly sound pollen, as viewed under the microscope. During 1866, it was freely and legitimately fertilized by other illegitimate plants belonging to the present and the following class, both of which include many highly fertile plants. Under these circumstances it yielded from eight capsules an average of 100·5 seeds, with a maximum of 123 and a minimum of 86; so that it produced 121 per cent of seeds in comparison with the normal standard. During 1864 it was allowed to be freely and legitimately fertilized by legitimate and illegitimate plants, and yielded an average, from eight capsules, of 104·2 seeds, with a maximum of 125 and a minimum of 90; consequently it exceeded the normal standard, producing 125 per cent of seeds. In this / case, as in some previous cases, pollen from legitimate plants added in a small degree to the fertility of the plant; and the fertility would, perhaps, have been still greater had not the summer of 1864 been very hot and certainly unfavourable to some of the plants of Lythrum.

Plant 23. This short-styled plant produced perfectly sound pollen. During 1866 it was freely and legitimately fertilized by the other illegitimate plants specified under the last experiment, and eight capsules yielded an average of 113·5 seeds, with a maximum of 123 and a minimum of 93. Hence this plant exceeded the normal standard, producing no less than 136 per cent of seeds.

Plant 24. This long-styled plant produced pollen which seemed under the microscope sound; but some of the grains did not swell when placed in water. During 1864 it was legitimately fertilized by legitimate and illegitimate plants in the same manner as Plant 22, but yielded an average, from ten capsules, of only 55 seeds, with a maximum of 88 and a minimum of 24, thus attaining 59 per cent of the normal fertility. This low degree of fertility, I presume, was owing to the unfavourable season; for during 1866, when legitimately fertilized by illegitimate plants in the manner described under No. 22, it yielded an average, from eight capsules, of 82 seeds, with a maximum of 120 and a minimum of 67, thus producing 88 per cent of the normal number of seeds.

Plant 25. The pollen of this long-styled plant contained a moderate number of poor and shrivelled grains; and this is a surprising circumstance, as it yielded an extraordinary number of seeds. During 1866 it was freely and legitimately fertilized by illegitimate plants, as described under No. 22, and yielded an average, from eight capsules, of 122·5 seeds, with a maximum of 149 and a minimum of 84. Hence this plant exceeded the normal standard, producing no less than 131 per cent of seeds.

CLASS VI

Illegitimate plants raised from mid-styled parents fertilized with pollen from the shortest stamens of the long-styled form

I raised from this union twenty-five plants, which proved to be seventeen long-styled and eight mid-styled, / but not one short-styled. None of these plants were in the least dwarfed. I examined, during the highly favourable season of 1866, the pollen of four plants; in one mid-styled plant, some of the anthers of the longest stamens were contabescent, but the pollen grains in the other anthers were mostly sound, as they were in all the anthers of the shortest stamens; in two other mid-styled and in one long-styled plant many of the pollen grains were small and shrivelled: and in the latter plant as many as a fifth or sixth part appeared to be in this state. I counted the seeds in five plants (Nos 26 to 30), of which two were moderately sterile and three fully fertile.

Plant 26. This mid-styled plant was freely and legitimately fertilized, during the rather unfavourable year 1864, by numerous surrounding legitimate and illegitimate plants. It yielded an average, from ten capsules, of 83·5 seeds, with a maximum of 110 and a minimum of 64, thus attaining 64 per cent of the normal fertility. During the highly favourable year 1866, it was freely and legitimately fertilized by illegitimate plants belonging to the present Class and to Class V, and yielded an average, from eight capsules, of 86 seeds, with a maximum of 109 and a minimum of 61, and thus attained 66 per cent of the normal fertility. This was the plant with some of the anthers of the longest stamens contabescent as above mentioned.

Plant 27. This mid-styled plant, fertilized during 1864 in the same manner as the last, yielded an average, from ten capsules, of 99·4 seeds, with a maximum of 122 and a minimum of 53, thus attaining to 76 per cent of the normal fertility. If the season had been more favourable, its fertility would probably have been somewhat greater, but, judging from the last experiment, only in a slight degree.

Plant 28. This mid-styled plant, when legitimately fertilized during the favourable season of 1866, in the manner described under No. 26, yielded an average, from eight capsules, of 89 seeds, with a maximum of 119 and a minimum of 69, thus producing 68 per cent of the full number of seeds. In the pollen of both sets of anthers, nearly as many grains were small and shrivelled as sound. /

Plant 29. This long-styled plant was legitimately fertilized during the unfavourable season of 1864, in the manner described under No. 26, and yielded an average, from ten capsules, of 84·6 seeds, with a maximum of 132 and a minimum of 47, thus attaining to 91 per cent of the normal fertility. During the highly favourable season of 1866, when fertilized in the manner described under No. 26, it yielded an average, from nine capsules (one poor capsule having been excluded), of 100 seeds, with a maximum of 121 and a minimum of 77. This plant thus exceeded the normal standard, and produced 107 per cent of seeds. In both sets of anthers there were a good many bad and shrivelled pollen grains, but not so many as in the last-described plant.

Plant 30. This long-styled plant was legitimately fertilized during 1866 in the manner described under No. 26, and yielded an average, from eight capsules, of 94 seeds, with a maximum of 106 and a minimum of 66; so that it exceeded the normal standard, yielding 101 per cent of seeds.

Plant 31. Some flowers on this long-styled plant were artificially and legitimately fertilized by one of its brother illegitimate mid-styled plants; and five capsules yielded an average of 90·6 seeds, with a maximum of 97 and a minimum of 79. Hence, as far as can be judged from so few capsules, this plant attained, under these favourable circumstances, 98 per cent of the normal standard.

CLASS VII

Illegitimate plants raised from mid-styled parents fertilized with pollen from the longest stamens of the short-styled form

It was shown in the last chapter that the union from which these illegitimate plants were raised is far more fertile than any other illegitimate union; for the mid-styled parent, when thus fertilized, yielded an average (all very poor capsules being excluded) of 102·8 seeds, with a maximum of 130; and the seedlings in the present class likewise have their fertility not at all lessened. Forty plants were raised; and these attained their full height and were covered with seed-capsules. / Nor did I observe any contabescent anthers. It deserves, also, particular notice that these plants, differently from what occurred in any of the previous classes, consisted of all three forms, namely, eighteen short-styled, fourteen long-styled, and eight mid-styled plants. As these plants were so fertile, I counted the seeds only in the two following cases.

Plant 32. This mid-styled plant was freely and legitimately fertilized during the unfavourable year of 1864, by numerous surrounding legitimate and illegitimate plants. Eight capsules yielded an average of 127·2 seeds, with a maximum of 144 and a minimum of 96; so that this plant attained 98 per cent of the normal standard.

Plant 33. This short-styled plant was fertilized in the same manner and at the same time with the last; and ten capsules yielded an average of 113·9, with a maximum of 137 and a minimum of 90. Hence this plant produced no less than 137 per cent of seeds in comparison with the normal standard.

Concluding remarks on the illegitimate offspring of the three forms of Lythrum salicaria

From the three forms occurring in approximately equal numbers in a state of nature, and from the results of sowing seed naturally produced, there is reason to believe that each form, when legitimately fertilized, reproduces all three forms in about equal numbers. Now, we have seen (and the fact is a very singular one) that the fifty-six plants produced from the long-styled form, illegitimately fertilized with pollen from the same form (Classes I and II), were all long-styled. The short-styled form, when self-fertilized (Class III), produced eight short-styled and one long-styled plant; and the mid-styled form, similarly treated (Class IV), produced three mid-styled and one long-styled offspring; so that these two forms, when illegitimately / fertilized with pollen from the same form, evince a strong, but not exclusive tendency to reproduce the parent form. When the short-styled form was illegitimately fertilized by the long-styled form (Class V), and again when the mid-styled was illegitimately fertilized by the long-styled (Class VI), in each case the two parent forms alone were reproduced. As thirty-seven plants were raised from these two unions, we may, with much confidence, believe that it is the rule that plants thus derived usually consist of both parent forms, but not of the third form. When, however, the mid-styled form was illegitimately fertilized by the longest stamens of the short-styled (Class VII), the same rule did not hold good; for the seedlings consisted of all three forms. The illegitimate union from which these latter seedlings were raised is, as previously stated, singularly fertile, and the seedlings themselves exhibited no signs of sterility and grew to their full height. From the consideration of these several facts, and from analogous ones to be given under Oxalis, it seems probable that in a state of nature the pistil of each form usually receives, through the agency of insects, pollen from the stamens of corresponding height from both the other forms. But the case last given shows that the application of two kinds of pollen is not indispensable for the production of all three forms. Hildebrand has suggested that the cause of all three forms being regularly and naturally reproduced, may be that some of the flowers are fertilized with one kind of pollen, and others on the same plant with the other kind of pollen. Finally, of the three forms, the long-styled evinces somewhat the strongest tendency to reappear among the offspring, whether both, or one, or neither of the parents are long-styled. /

The lessened fertility of most of these illegitimate plants is in many respects a highly remarkable phenomenon. Thirty-three plants in the seven classes were subjected to various trials, and the seeds carefully counted. Some of them were artificially fertilized, but the far greater number were freely fertilized (and this is the better and natural plan) through the agency of insects, by other illegitimate plants. In the right hand, or percentage column, in Table 30, a wide difference in fertility between the plants in the first four and the last three classes may be perceived. In the first four classes the plants are descended from the three forms illegitimately fertilized with pollen taken from the same form, but only rarely from the same plant. It is necessary to observe this latter circumstance; for, as

TABLE 30

Tabulated results of the fertility of the foregoing illegitimate plants, when legitimately fertilized, generally by illegitimate plants, as described under each experiment. Plants 11, 12, and 13 are excluded, as they were illegitimately fertilized

Normal standard of fertility of the three forms, when legitimately and naturally fertilized

Form	Average number of seeds per capsule	Maximum number in any one capsule	Minimum number in any one capsule
Long-styled	93	159	No record was kept, as all
Mid-styled	130	151	very poor capsules were
Short-styled	83·5	112	rejected.

TABLE 30 – *continued*

CLASSES I AND II

Illegitimate plants raised from long-styled parents fertilized with pollen from one-form mid-length or shortest stamens

Number of plant	Form	Average number of seeds per capsule	Maximum number in any one capsule	Minimum number in any one capsule	Average number of seeds, expressed as the percentage of the normal standard
Plant 1	Long-styled	0	0	0	0
Plant 2	,,	4·5	?	0	5
Plant 3	,,	4·5	?	0	5
Plant 4	,,	4·5	?	0	5
Plant 5	,,	0 or 1	2	0	0 or 1
Plant 6	,,	0	0	0	0
Plant 7	,,	36·1	47	22	39
Plant 8	,,	41·1	73	11	44
Plant 9	,,	57·1	86	23	61
Plant 10	,,	44·2	69	25	47

CLASS III

Illegitimate plants raised from short-styled parents fertilized with pollen from own-form shortest stamens

Plant 14	Short-styled	28·3	51	11	33
Plant 15	,,	32·6	49	20	38
Plant 16	,,	77·8	97	60	94
Plant 17	Long-styled	76·3	88	57	82 /

CLASS IV

Illegitimate plants raised from mid-styled parents fertilized with pollen from own-form longest stamens

Plant 18	Mid-styled	102·6	131	63	80
Plant 19	,,	73·4	87	64	56
Plant 20	Long-styled	69·6	83	52	75

CLASS V

·*Illegitimate plants raised from short-styled parents fertilized with pollen from the mid-length stamens of the long-styled form*

Plant 21	Short-styled	43·0	63	26	52
Plant 22	,,	100·5	123	86	121
Plant 23	,,	113·5	123	93	136
Plant 24	Long-styled	82·0	120	67	88
Plant 25	,,	122·5	149	84	131

TABLE 30 – *continued*

CLASS VI

Illegitimate plants raised from mid-styled parents fertilized with pollen from the shortest stamens of the long-styled form

Number of plant	Form	Average number of seeds per capsule	Maximum number in any one capsule	Minimum number in any one capsule	Average number of seeds, expressed as the percentage of the normal standard
Plant 26	Mid-styled	86·0	109	61	66
Plant 27	„	99·4	122	53	76
Plant 28	„	89·0	119	69	68
Plant 29	Long-styled	100·0	121	77	107
Plant 30	„	94·0	106	66	101
Plant 31	„	90·6	97	79	98

CLASS VII

Illegitimate plants raised from mid-styled parents fertilized with pollen from the longest stamens of the short-styled form

| Plant 32 | Mid-styled | 127·2 | 144 | 96 | 98 |
| Plant 33 | Short-styled | 113·9 | 137 | 90 | 137 |

I have elsewhere shown,[3] most plants, when fertilized with their own pollen, or that from the same plant, are in some degree sterile, and the seedlings raised from such unions are likewise in some degree sterile, dwarfed, and feeble. None of the nineteen illegitimate plants in the first four classes were completely fertile; one, however, was nearly so, yielding 96 per cent of the proper number of seeds. From this high degree of fertility we have many descending gradations, till we reach an absolute zero, when the plants, though bearing many flowers, did not produce, during successive years, a single seed or even seed-capsule. Some of the most sterile plants did not even yield a single seed when legitimately fertilized with pollen from legitimate plants. There is good reason to believe that the first seven plants in Classes I and II were the offspring / of a long-styled plant fertilized with pollen from its own-form shortest stamens, and these plants were the most sterile of all. The remaining plants in Classes I and II were almost certainly the product of pollen from the mid-length stamens, and although very sterile, they were less so than the first set. None of the plants in the first

[3] *The Effects of Cross and Self-fertilisation in the Vegetable Kingdom*, 1876.

four classes attained their full and proper stature; the first seven, which were the most sterile of all (as already stated), were by far the most dwarfed, several of them were never reaching to half their proper height. These same plants did not flower at so early an age, or at so early a period in the season, as they ought to have done. The anthers in many of their flowers, and in the flowers of some other plants in the first six classes, were either contabescent or included numerous small and shrivelled pollen grains. As the suspicion at one time occurred to me that the lessened fertility of the illegitimate plants might be due to the pollen alone having been affected, I may remark that this certainly was not the case; for several of them, when fertilized by sound pollen from legitimate plants, did not yield the full complement of seeds; hence it is certain that both the female and male reproductive organs were affected. In each of the seven classes, the plants, though descended from the same parents, sown at the same time and in the same soil, differed much in their average degree of fertility.

Turning now to the fifth, sixth, and seventh classes, and looking to the right-hand column of the table, we find nearly as many plants with a percentage of seeds above the normal standard as beneath it. As with most plants the number of seeds produced varies much, it might be thought that the present case was one merely of variability. But this view must be rejected, / as far as the less fertile plants in these three classes are concerned: first, because none of the plants in Class V attained their proper height, which shows that they were in some manner affected; and secondly, because many of the plants in Classes V and VI produced anthers which were either contabescent or included small and shrivelled pollen grains. And as in these cases the male organs were manifestly deteriorated, it is by far the most probable conclusion that the female organs were in some cases likewise affected, and that this was the cause of the reduced number of seeds.

With respect to the six plants in these three classes which yielded a very high percentage of seeds, the thought naturally arises that the normal standard of fertility for the long-styled and short-styled forms (with which alone we are here concerned) may have been fixed too low, and that the six illegitimate plants are merely fully fertile. The standard for the long-styled form was deduced by counting the seeds in twenty-three capsules, and for the short-styled form from twenty-five capsules. I do not pretend that this is a sufficient number of capsules for absolute accuracy; but my experience has led me to believe that a very fair result may thus be gained. As, however, the maximum number observed in the twenty-five capsules of the short-styled form was low, the standard in this case may possibly be not quite high enough. But it should be observed, in the case of the illegitimate plants, that in order to avoid over-estimating their infertility, ten very fine capsules were always selected; and the years 1865 and 1866, during which the plants in the three latter classes were experimented on, were highly favourable for seed production. Now, if this / plan of selecting very fine capsules during favourable seasons had been followed for obtaining the normal standards, instead of taking, during various seasons, the first capsules which came to hand, the standards would undoubtedly have been considerably higher; and thus the fact of the six foregoing plants appearing to yield an unnaturally high percentage of seeds may, perhaps, be explained. On this view, these plants are, in fact, merely fully fertile, and not fertile to an abnormal degree. Nevertheless, as characters of all kinds are liable to variation, especially with organisms unnaturally treated, and as in the four first and more sterile classes, the plants derived from the same parents and treated in the same manner, certainly did vary much in sterility, it is possible that certain plants in the latter and more fertile classes may have varied so as to have acquired an abnormal degree of fertility. But it should be noticed that, if my standards err in being too low, the sterility of all the many sterile plants in the several classes will have to be estimated by so much the higher. Finally, we see that the illegitimate plants in the four first classes are all more or less sterile, some being absolutely barren, with one alone almost completely fertile; in the three latter classes, some of the plants are moderately sterile, whilst others are fully fertile, or possibly fertile in excess.

The last point which need here be noticed is that, as far as the means of comparison serve, some degree of relationship generally exists between the infertility of the illegitimate union of the several parent forms and that of their illegitimate offspring. Thus the two illegitimate unions, from which the plants in Classes VI and VII were derived, yielded a fair amount of seed, and only a few of these plants are in / any degree sterile. On the other hand, the illegitimate unions between plants of the same form always yield very few seeds, and their seedlings are very sterile. Long-styled parent forms when fertilized with pollen from their own-form shortest stamens appear to be rather more sterile than when fertilized with their own-form mid-length stamens; and the seedlings from the former union were much more sterile than those from the latter union. In opposition to this relationship, short-styled plants illegitimately fertilized with pollen from the mid-length stamens of the long-styled form (Class V) are very sterile; whereas some of the offspring raised from this union were far from being highly sterile. It may be added that there is a tolerably close parallelism in all the classes between the degree of sterility of the plants and their dwarfed stature. As previously stated, an illegitimate plant fertilized with pollen from a legitimate plant has its fertility slightly increased. The importance of the several foregoing conclusions will be apparent at the close of this chapter, when the illegitimate unions between the forms of the same species and their illegitimate offspring, are compared with the hybrid unions of distinct species and their hybrid offspring.

OXALIS

No one has compared the legitimate and illegitimate offspring of any trimorphic species in this genus. Hildebrand sowed illegitimately fertilized seeds of *Oxalis Valdiviana*,[4] but they did not germinate; and this fact, as he remarks, supports my view that an illegitimate union resembles a hybrid one between / two distinct species, for the seeds in this latter case are often incapable of germination.

The following observations relate to the nature of the form which appear among the legitimate seedlings of *Oxalis Valdiviana*, Hildebrand raised, as described in the paper just referred to, 211 seedlings from all six legitimate unions, and the three forms appeared among the offspring from each union. For instance, long-styled plants were legitimately fertilized with pollen from the longest stamens of the mid-styled form, and the seedlings consisted of 15 long-styled, 18 mid-styled, and 6 short-styled. We here see that a few short-styled plants were produced, though neither parent was short-styled; and so it was with the other legitimate unions. Out of the above 211 seedlings, 173 belonged to the same two forms as their parents, and only 38 belonged to the third form distinct from either parent. In the case of *O. Regnelli*, the result, as observed by Hildebrand, was nearly the same, but more striking; all the offspring from four of the legitimate unions consisted of the two parent forms, whilst among the seedlings from the other two legitimate unions the third form appeared. Thus, of the 35 seedlings from the six legitimate unions, 35 belonged to the same two forms as their parents, and only 8 to the third form. Fritz Müller also raised in Brazil seedlings from long-styled plants of *O. Regnelli*

---

[4] *Bot. Zeitung*, 1871, p. 433, footnote.

legitimately fertilized with pollen from the longest stamens of the mid-styled form, and all these belonged to the two parent forms.[5] Lastly, seedlings were raised by me from long-styled plants of *O. speciosa* legitimately fertilized by the short-styled form, and from the latter reciprocally fertilized by the long-styled; and these consisted of 33 long-styled and 26 short-styled plants with not one mid-styled form. There can, therefore, be no doubt that the legitimate offspring from any two forms of Oxalis tend to belong to the same two forms as their parents; but that a few seedlings belonging to the third form occasionally make their appearance; and this latter fact, as Hildebrand remarks, may be attributed to atavism, as some of their progenitors will almost certainly have belonged to the third form.

When, however, any form of Oxalis is fertilized illegitimately / with pollen from the same form, the seedlings appear to belong invariably to this form. Thus Hildebrand states[6] that long-styled plants of *O. rosea* growing by themselves have been propagated in Germany year after year by seed, and have always produced long-styled plants. Again, 17 seedlings were raised from mid-styled plants of *O. hedysaroides* growing by themselves, and these were all mid-styled. So that the forms of Oxalis, when illegitimately fertilized with their own pollen, behave like the long-styled form of *Lythrum salicaria*, which when thus fertilized always produced with me long-styled offspring.

PRIMULA

PRIMULA SINENSIS

I raised during February, 1862, from some long-styled plants illegitimately fertilized with pollen from the same form, twenty-seven seedlings. These were all long-styled. They proved fully fertile or even fertile in excess; for ten flowers, fertilized with pollen from other plants of the same lot, yielded nine capsules, containing on an average 39·75 seeds, with a maximum in one capsule of 66 seeds. Four other flowers legitimately crossed with pollen from a legitimate plant, and four flowers on the latter crossed with pollen from the illegitimate seedlings, yielded seven capsules with an average of 53 seeds, with a maximum of 72. I must here state that I have found some difficulty in estimating the normal standard of fertility for the several unions of this species, as the results differ much during successive years, and the seeds vary so greatly in size that it is hard to / decide which ought to be considered good. In order to avoid over-estimating the infertility of

---

[5] *Jenaische Zeitschrift*, etc., vol. vi, 1871, p. 75.
[6] *Ueber den Trimorphismus in der Gattung Oxalis: Monatsberichte der Akad. der Wissen. zu Berlin*, 21 June, 1866, p. 373 and *Bot. Zeitung*, 1871, p. 435.

the several illegitimate unions, I have taken the normal standard as low as possible.

From the foregoing twenty-seven illegitimate plants, fertilized with their own-form pollen, twenty-five seedling grandchildren were raised, and these were all long-styled; so that from the two illegitimate generations fifty-two plants were raised, and all without exception proved long-styled. These grandchildren grew vigorously, and soon exceeded in height two other lots of illegitimate seedlings of different parentage and one lot of equal-styled seedlings presently to be described. Hence I expected that they would have turned out highly ornamental plants; but when they flowered, they seemed, as my gardener remarked, to have gone back to the wild state; for the petals were pale-coloured, narrow, sometimes not touching each other, flat, generally deeply notched in the middle, but not flexuous on the margin, and with the yellow eye or centre conspicuous. Altogether these flowers were strikingly different from those of their progenitors; and this, I think, can only be accounted for on the principle of reversion. Most of the anthers on one plant were contabescent. Seventeen flowers on the grandchildren were illegitimately fertilized with pollen taken from other seedlings of the same lot, and produced fourteen capsules, containing on an average 29·2 seeds; but they ought to have contained about 35 seeds. Fifteen flowers legitimately fertilized with pollen from an illegitimate short-styled plant (belonging to the lot next to be described) produced fourteen capsules, containing an average of 46 seeds; they ought to have contained at least 50 seeds. Hence these grandchildren of illegitimate descent appear / to have lost, though only in a very slight degree, their full fertility.

We will now turn to the short-styled form: from a plant of this kind, fertilized with its own-form pollen, I raised, during February, 1862, eight seedlings, seven of which were short-styled and one long-styled. They grew slowly, and never attained to the full stature of ordinary plants; some of them flowered precociously, and others late in the season. Four flowers on these short-styled seedlings and four on the one long-styled seedling were illegitimately fertilized with their own-form pollen and produced only three capsules, containing on an average 23·6 seeds, with a maximum of 29; but we cannot judge of their fertility from so few capsules: and I have greater doubts about the normal standard for this union than about any other; but I believe that rather above 25 seeds would be a fair estimate. Eight flowers on these same short-styled plants, and the one long-styled illegitimate plant were reciprocally and legitimately crosed; they produced five capsules, which contained an average of 28·6 seeds, with a maximum of 36. A reciprocal cross between legitimate plants of the two forms would have yielded an average of at least 57 seeds, with a possible maximum of 74 seeds; so that these illegitimate plants were sterile when legitimately crossed.

I succeeded in raising from the above seven short-styled illegitimate plants, fertilized with their own-form pollen, only six plants – grandchildren of the first union. These, like their parents, were of low stature, and had so poor a constitution that four died before flowering. With ordinary plants it has been a rare event with me to have more than a single plant die out of a large lot. The two grandchildren which / lived and flowered were short-styled; and twelve of their flowers were fertilized with their own-form pollen and produced twelve capsules containing an averge of 28·2 seeds; so that these two plants, though belonging to so weakly a set, were rather more fertile than their parents, and perhaps not in any degree sterile. Four flowers on the same two grandchildren were legitimately fertilized by a long-styled illegitimate plant, and produced four capsules, containing only 32·2 seeds instead of about 64 seeds, which is the normal average for legitimate short-styled plants legitimately crossed.

By looking back, it will be seen that I raised at first from a short-styled plant fertilized with its own-form pollen one long-styled and seven short-styled illegitimate seedlings. These seedlings were legitimately intercrossed, and from their seed fifteen plants were raised, grandchildren of the first illegitimate union, and to my surprise all proved short-styled. Twelve short-styled flowers borne by these grandchildren were illegitimately fertilized with pollen taken from other plants of the same lot, and produced eight capsules which contained an average of 21·8 seeds, with a maximum of 35. These figures are rather below the normal standard for such a union. Six flowers were also legitimately fertilized with pollen from an illegitimate long-styled plant and produced only three capsules, containing on an average 23·6 seeds, with a maximum of 35. Such a union in the case of a legitimate plant ought to have yielded an average of 64 seeds, with a possible maximum of 73 seeds.

*

Summary on the transmission of form, constitution, and fertility of the
illegitimate offspring of Primula Sinensis

In regard to the long-styled plants, their / illegitimate offspring, of which fifty-two were raised in the course of two generations, were all long-styled.[7] These plants grew vigorously; but the flowers in one instance were small, appearing as if they had reverted to the wild state. In the first illegitimate generation they were perfectly fertile, and in the second their fertility was only very slightly impaired. With respect to the short-styled plants, twenty-four out of twenty-five of their illegitimate offspring were short-styled. They were dwarfed in stature, and one lot of grandchildren had so poor a constitution that four out of six plants perished before flowering. The two survivors, when illegitimately fertilized with their own-form pollen, were rather less fertile than they ought to have been; but their loss of fertility was clearly shown in a special and unexpected manner, namely, when legitimately fertilized by other illegitimate plants: thus altogether eighteen flowers were fertilized in this manner, and yielded twelve capsules, which included on an average only 28·5 seeds, with a maximum of 45. Now a legitimate short-styled plant would have yielded, when legitimately fertilized, an average of 64 seeds, with a possible maximum of 74. This particular kind of infertility will perhaps be best appreciated by a simile: we may assume that with mankind six children would be born on an average from an ordinary marriage; but that only three would be born from an incestuous marriage. According to the analogy of *Primula Sinensis*, the children of such / incestuous marriages, if they continued to marry incestuously, would have their sterility only slightly increased; but their fertility would not be restored by a proper marriage; for if two children, both of incestuous origin, but in no degree related to each other, were to marry, the marriage would of course be strictly legitimate, nevertheless they would not give birth to more than half the full and proper number of children.

Equal-styled variety of Primula Sinensis. As any variation in the structure of the reproductive organs, combined with changed function, is a rare event, the

---

[7] Dr Hildebrand, who first called attention to this subject (*Bot. Zeitung*, 1864, p. 5), raised from a similar illegitimate union seventeen plants, of which fourteen were long-styled and three short-styled. From a short-styled plant illegitimately fertilized with its own pollen he raised fourteen plants, of which eleven were short-styled and three long-styled.

following cases are worth giving in detail. My attention was first called to the subject by observing, in 1862, a long-styled plant, descended from a self-fertilized long-styled parent, which had some of its flowers in an anomalous state, namely, with the stamens placed low down in the corolla as in the ordinary long-styled form, but with the pistils so short that the stigmas stood on a level with the anthers. These stigmas were nearly as globular and as smooth as in the short-styled form, instead of being elongated and rough as in the long-styled form. Here, then, we have combined in the same flower, the short stamens of the long-styled form with a pistil closely resembling that of the short-styled form. But the structure varied much even on the same umbel: for in two flowers the pistil was intermediate in length between that of the long- and that of the short-styled form, with the stigma elongated as in the former, and smooth as in the latter; and in three other flowers the structure was in all respects like that of the long-styled form. These modifications appeared to me so remarkable that I fertilized eight of the flowers with their own pollen, and obtained five capsules, which contained on an average 43 seeds; and this number shows that the flowers had become abnormally fertile in comparison with those of ordinary long-styled plants when self-fertilized. I was thus led to examine the plants in several small collections, and the result showed that the equal-styled variety was not rare.

TABLE 31   *Primula Sinensis*

Name of owner or place	Long-styled form	Short-styled form	Equal-styled variety
Mr Horwood	0	0	17
Mr Duck	20	0	9
Baston	30	18	15
Chichester	12	9	2
Holwood	42	12	0
High Elms	16	0	0
Westerham	1	5	0
My own plants fro n purchased seeds	13	7	0
Total	134	51	43

In a state of nature the long- and short-styled forms would no doubt occur in nearly equal numbers, as I infer from the analogy of the other heterostyled species of Primula, and from having / raised the two forms of the present species in exactly the same number from flowers which had been *legitimately* crossed. The preponderance in the above table of the long-styled form over the short-styled (in the proportion of 134 to 51) results from gardeners generally collecting seed from self-fertilized flowers; and the long-styled flowers produce spontaneously much more seed (as shown in the first chapter) than the short-styled, owing to the anthers of the long-styled form being placed low down in the corolla, so that, when the flowers fall off, the anthers are dragged over the stigma; and we now also know that long-styled plants, when self-fertilized, very generally reproduce long-styled offspring. From the consideration of this table, it occurred to me in the year 1862, that almost all the plants of the Chinese primrose cultivated in England would sooner or later become long-styled or equal-styled; and now, at the close of 1876, I have had five small collections of plants examined, and almost all consisted of long-styled, with some more or less well-characterized equal-styled plants, but with not one short-styled.

With respect to the equal-styled plants in the table, Mr Horwood raised from purchased seeds four plants, which he remembered were certainly not long-styled, but either short- or equal-styled, probable the latter. These four plants were kept separate and allowed to fertilize themselves; from their seed the seventeen plants in the table were raised, all of which proved equal-styled. The stamens stood low down in the corolla as in the long-styled form; and the stigmas, which were globular and / smooth, were either completely surrounded by the anthers, or stood close above them. My son William made drawings for me, by the aid of the camera, of the pollen of one of the above equal-styled plants; and, in accordance with the position of the stamens, the grains resembled in their small size those of the long-styled form. He also examined pollen from two equal-styled plants at Southampton: and in both of them the grains differed extremely in size in the same anthers, a large number being small and shrivelled, whilst many were fully as large as those of the short-styled form and rather more globular. It is probable that the large size of these grains was due, not to their having assumed the character of the short-styled form, but to monstrosity; for Max Wichura has observed pollen grains of monstrous size in certain hybrids. The vast number of the small shrivelled grains in the above two cases explains the fact that, though equal-styled plants are generally fertile in a high degree, yet some of them yield few seeds. I may add that my son compared, in 1875, the grains from two white-flowered plants, in both of which the pistil projected above the anthers, but neither were properly long-styled or equal-styled; and in the one in which the stigma projected most, the grains were in diameter to those in the other plant, in which the stigma projected less, as 100 to 88; whereas the difference between the grains from perfectly characterized long-styled and short-styled plants is as 100 to 57. So that these two plants were in an intermediate condition. To return to the 17 plants in the first line of Table 31: from the relative position of their stigmas and anthers, they could hardly fail to fertilize themselves: and accordingly four of them spontaneously yielded no less than 180 capsules; of these Mr Horwood selected eight fine capsules for sowing; and they included on an average 54·8 seeds, with a maximum of 72. He gave me thirty other capsules, taken by hazard, of which twenty-seven contained good seeds, averaging 35·5, with a maximum of 70: but if six poor capsules, each with less than 13 seeds, be excluded, the average rises to 42·5. These are higher numbers than could be expected from either well-characterized form if self-fertilized; and this high degree of fertility accords with the view that the male organs belonged to one form, and the female organs partially to the other form; so that a self-union in the case of the equal-styled variety is in fact a legitimate union.

The seed saved from the above seventeen self-fertilized equal-styled / plants produced sixteen plants, which all proved equal-styled, and resembled their parents in all the above-specified respects. The stamens, however, in one plant were seated higher up the tube of the corolla than in the true long-styled form; in another plant almost all the anthers were contabescent. These sixteen plants were the grandchildren of the four original plants which it is believed were equal-styled; so that this abnormal condition was faithfully transmitted, probably through three, and certainly through two generations. The fertility of one of these grandchildren was carefully observed; six flowers were fertilized with pollen from the same flower, and produced six capsules, containing on an average 68 seeds, with a maximum of 82, and a minimum of 40. Thirteen capsules spontaneously self-fertilized yielded an average of 53·2 seeds, with the astonishing maximum in one of 97 seeds. In no legitimate union has so high an average as 68 seeds been observed by me, or nearly so high a maximum as 82 and 97. These plants, therefore, not only have lost their proper heterostyled structure and peculiar functional powers, but have acquired an abnormal grade of fertility – unless, indeed, their high fertility may be accounted for by the stigmas receiving pollen from the circumjacent anthers at exactly the most favourable period.

With respect to Mr Duck's lot in Table 31, seed was saved from a single plant, of which the form was not observed, and this produced nine equal-styled and twenty long-styled plants. The equal-styled resembled in all respects those previously described: and eight of their capsules spontaneously self-fertilized contained on an average 44·4 seeds, with a maximum of 61 and a minimum of 23. In regard to the twenty long-styled plants, the pistil in some of the flowers did not project quite so high as in ordinary long-styled flowers; and the stigmas, though properly elongated, were smooth, so that we have here a slight approach in structure to the pistil of the short-styled form. Some of these long-styled plants also approached the equal-styled in function; for one of them produced no less than fifteen spontaneously self-fertilized capsules, and of these eight contained, on an average, 31·7 seeds, with a maximum of 61. This average would be rather low for a long-styled plant artificially fertilized with its own pollen, but is high for one spontaneously self-fertilized. For instance, thirty-four capsules produced by the illegitimate grandchildren of a long-styled plant, spontaneously self-fertilized, contained / on an average only 9·1 seeds, with a maximum of 46. Some seeds indiscriminately saved from the foregoing twenty-nine equal-styled and long-styled plants produced sixteen seedlings; grandchildren of the original plant belonging to Mr Duck; and these consisted of fourteen equal-styled and two long-styled plants; and I mention this fact as an additional instance of the transmission of the equal-styled variety.

The third lot in the table, namely the Baston plants, are the last which need be mentioned. The long- and short-styled plants, and the fifteen equal-styled plants, were descended from two distinct stocks. The latter were derived from a single plant, which the gardener is positive was not long-styled; hence probably, it was equal-styled. In all these fifteen plants the anthers occupying the same position as in the long-styled form, closely surrounded the stigma, which in one instance alone was slightly elongated. Notwithstanding this position of the stigma, the flowers, as the gardener assured me, did not yield many seeds; and this difference from the foregoing cases may perhaps have been caused by the pollen being bad, as in some of the Southampton equal-styled plants.

Conclusions with respect to the equal-styled variety of P. Sinensis

That this is a variation, and not a third or distinct form, as in the tri-morphic genera Lythrum and Oxalis, is clear; for we have seen its first appearance in one out of a lot of illegitimate long-styled plants; and in the case of Mr Duck's seedlings, long-styled plants, only slightly deviat-ing from the normal state, as well as equal-styled plants were produced from the same self-fertilized parent. The position of the stamens in their proper place low down in the tube of the corolla, together with the small size of the pollen grains, show, first, that the equal-styled variety is a modification of the long-styled form, and, secondly, that the pistil is the part which has varied most, as indeed was obvious in many of the plants. This variation is of frequent occurrence, and is strongly inherited when it has once appeared. It would, however, have possessed / little interest if it had consisted of a mere change of structure; but this is accompanied by modified fertility. Its occurrence apparently stands in close relation with the illegitimate birth of the parent plant; but to this whole subject I shall hereafter recur.

PRIMULA AURICULA

Although I made no experiments on the illegitimate offspring of this species, I refer to it for two reasons: First, because I have observed two equal-styled plants in which the pistil resembled in all respects that of the long-styled form, whilst the stamens had become elongated as in the short-styled form, so that the stigma was almost surrounded by the anthers. The pollen grains, however, of the elongated stamens resembled in their small size those of the shorter stamens proper to the long-styled form. Hence these plants have become equal-styled by the increased length of the stamens, instead of, as with *P. Sinensis* by the diminished•length of the pistil. Mr J. Scott observed five other plants in the same state, and he shows[8] that one of them, when self-fertilized, yielded more seed than an ordinary long- or short-styled form would have done when similarly fertilized, but that it was far inferior in fertility to either form when legitimately crossed. Hence it appears that the male and female organs of this equal-styled variety have been modified in some special manner, not only in structure, but in functional powers. This, moreover, is shown by the singular fact that both the long-styled and short-styled plants, fertilized with pollen from the equal-styled variety, yield a lower average of seed than when these two forms are fertilized with their own pollen.

The second point which deserves notice is that florists always throw away the long-styled plants, and save seed exclusively from the short-styled form.

[8] *Journal Proc. Linn. Soc.*, viii (1864), p. 91.

Nevertheless, as Mr Scott was informed by a man who raises this species extensively in Scotland, about one-fourth of the seedlings appear long-styled; so that the short-styled form of the Auricula, when fertilized by its own pollen, does not reproduce the same form in so large a proportion as in the case of *P. Sinensis*. We may further infer / that the short-styled form is not rendered quite sterile by a long course of fertilization with pollen of the same form; but as there would always be some liability to an occasional cross with the other form, we cannot tell how long self-fertilization has been continued.

PRIMULA FARINOSA

Mr Scott says[9] that it is not at all uncommon to find equal-styled plants of this heterostyled species. Judging from the size of the pollen grains, these plants owe their structure, as in the case of *P. auricula*, to the abnormal elongation of the stamens of the long-styled form. In accordance with this view, they yield less seed when crossed with the long-styled form than with the short-styled. But they differ in an anomalous manner from the equal-styled plants of *P. auricula* in being extremely sterile with their own pollen.

PRIMULA ELATIOR

It was shown in the first chapter, on the authority of Herr Breitenbach, that equal-styled flowers are occasionally found on this species whilst growing in a state of nature; and this is the only instance of such an occurrence known to me, with the exception of some wild plants of the Oxlip – a hybrid between *P. veris* and *vulgaris* – which were equal-styled. Herr Breitenbach's case is remarkable in another way; for equal-styled flowers were found in two instances on plants which bore both long-styled and short-styled flowers. In every other instance these two forms and the equal-styled variety have been produced by distinct plants.

Primula vulgaris, Brit. Fl.
Var. *acaulis* of Linn. and *P. acaulis* of Jacq.

Var. *rubra*. – Mr Scott states[10] that this variety, which grew in the Botanic Garden in Edinburgh, was quite sterile when fertilized with pollen from the common primrose, as well as from a white variety of the same / species, but that some of the plants, when artificially fertilized with their own pollen, yielded a moderate supply of seed. He was so kind as to send me some of these self-fertilized seeds, from

[9] *Journal Proc. Linn. Soc.*, viii (1864), p. 115.
[10] Ibid., p. 98.

which I raised the plants immediately to be described. I may premise that the results of my experiments on the seedlings, made on a large scale, do not accord with those by Mr Scott on the parent plant.

First, in regard to the transmission of form and colour. The parent plant was long-styled, and of a rich purple colour. From the self-fertilized seed 23 plants were raised; of these 18 were purple of different shades, with 2 of them a little streaked and freckled with yellow, thus showing a tendency to reversion; and 5 were yellow, but generally with a brighter orange centre than in the wild flower. All the plants were profuse flowerers. All were long-styled; but the pistil varied a good deal in length even on the same plant, being rather shorter, or considerably longer, than in the normal long-styled form; and the stigmas likewise varied in shape. It is, therefore, probable that an equal-styled variety of the primrose might be found on careful search; and I have received two accounts of plants apparently in this condition. The stamens always occupied their proper position low down in the corolla; and the pollen grains were of the small size proper to the long-styled form, but were mingled with many minute and shrivelled grains. The yellow-flowered and the purple-flowered plants of this first generation were fertilized under a net with their own pollen, and the seed separately sown. From the former, 22 plants were raised and all were yellow and long-styled. From the latter or the purple-flowered plants, 24 long-styled plants were raised, of which 17 were purple and 7 yellow. / In this last case we have an instance of reversion in colour, without the possibility of any cross, to the grandparents or more distinct progenitors of the plants in question. Altogether 23 plants in the first generation and 46 in the second generation were raised; and the whole of these 69 illegitimate plants were long-styled!

Eight purple-flowered and two yellow-flowered plants of the first illegitimate generation were fertilized in various ways with their own pollen and with that of the common primrose; and the seeds were separately counted, but as I could detect no difference in fertility between the purple and yellow varieties, the results are run together in Table 32. (See next page.)

If we compare the figures in this table with those given in the first chapter, showing the normal fertility of the common primrose, we shall see that the illegitimate purple- and yellow-flowered varieties are very sterile. For instance, 72 flowers were fertilized with their own pollen and produced only 11 good capsules; but by the standard they ought to have produced 48 capsules; and each of these ought to have

TABLE 32   *Primula vulgaris*

Nature of plant experimented on, and kind of union	Number of flowers fertilized	Number of capsules produced	Average number of seeds per capsule	Maximum number of seeds in any one capsule	Minimum number of seeds in any one capsule
Purple- and yellow-flowered illegitimate long-styled plants, *illegitimately* fertilized with pollen from the same plant	72	11	11·5	26	5
Purple- and yellow-flowered illegitimate long-styled plants, *illegitimately* fertilised with pollen from the common long-styled primrose	72	39	31·4	62	3
Or, if the ten poorest capsules, including less than 15 seeds, be rejected, we get	72	29	40·6	62	18
Purple- and yellow-flowered illegitimate long-styled plants, *legitimately* fertilized with pollen from the common short-styled primrose	26	18	36·4	60	9
Or, if the two poorest capsules, including less than 15 seeds, be rejected, we get	26	16	41·2	60	15
The long-styled form of the common primrose, *illegitimately* fertilized with pollen from the long-styled illegitimate purple- and yellow-flowered plants	20	14	15·4	46	1
Or, if the three poorest capsules be rejected, we get	20	11	18·9	46	8
The short-styled form of the common primrose, *legitimately* fertilized with pollen from the long-styled illegitimate purple- and yellow-flowered plants	10	6	30·5	61	6

contained on an average 52·2 seeds, instead of only 11·5 seeds. When these plants were illegitimately and legitimately fertilized with pollen from the common primrose, the average numbers were increased, but were far from attaining the normal standards. So it was when both forms of the common primrose were fertilized with pollen from these illegitimate plants; and this shows that their male as well as their female organs were in a deteriorated condition. The sterility of these

plants was shown in another way, namely, by their not producing any
capsules when the access of all insects (except such minute ones as
Thrips) was prevented; for under these circumstances the common
long-styled / primrose produces a considerable number of capsules.
There can, therefore, be no doubt that the fertility of / these plants was
greatly impaired. The loss is not correlated with the colour of the
flower; and it was to ascertain this point that I made so many experiments. As the parent plant growing in Edinburgh was found by Mr
Scott to be in a high degree sterile, it may have transmitted a similar
tendency to its offspring, independently of their illegitimate birth. I
am, however, inclined to attribute some weight to the illegitimacy of
their descent, both from the analogy of other cases, and more especially from the fact that when the plants were *legitimately* fertilized with
pollen of the common primrose they yielded an average, as may be
seen in the table, of only 5 more seeds than when *illegitimately* fertilized
with the same pollen. Now we know that it is eminently characteristic
of the illegitimate offspring of *Primula Sinensis* that they yield but few
more seeds when legitimately fertilized than when fertilized with their
own-form pollen.

Primula veris, Brit. Fl.
Var. *officinalis* of Linn., *P. officinalis* of Jacq.

Seeds from the short-styled form of the cowslip fertilized with pollen
from the same form germinate so badly that I raised from three
successive sowings only fourteen plants, which consisted of nine short-
styled and five long-styled plants. Hence the short-styled form of the
cowslip, when self-fertilized, does not transmit the same form nearly so
truly as does that of *P. Sinensis*. From the long-styled form, always
fertilized with its own-form pollen, I raised in the first generation three
long-styled plants – from their seed 53 long-styled grandchildren –
from their seed 4 long-styled great-grandchildren – from their seed 20
long-styled great-great-grandchildren – and lastly, / from their seed 8
long-styled and 2 short-styled great-great-great-grandchildren. In this
last generation short-styled plants appeared for the first time in the
course of the six generations – the parent long-styled plant which was
fertilized with pollen from another plant of the same form being
counted as the first generation. Their appearance may be attributed to
atavism. From two other long-styled plants, fertilized with their own-
form pollen, 72 plants were raised, which consisted of 68 long-styled

and 4 short-styled. So that altogether 162 plants were raised from illegitimately fertilized long-styled cowslips, and these consisted of 156 long-styled and 6 short-styled plants.

We will now turn to the fertility and powers of growth possessed by the illegitimate plants. From a short-styled plant, fertilized with its own-form pollen, one short-styled and two long-styled plants, and from a long-styled plant similarly fertilized three long-styled plants were at first raised. The fertility of these six illegitimate plants was carefully observed; but I must premise that I cannot give any satisfactory standard of comparison as far as the number of the seeds is concerned; for though I counted the seeds of many legitimate plants fertilized legitimately and illegitimately, the number varied so greatly during successive seasons that no one standard will serve well for illegitimate unions made during different seasons. Moreover the seeds in the same capsule frequently differ so much in size that it is scarcely possible to decide which ought to be counted as good seed. There remains as the best standard of comparison the proportional number of fertilized flowers which produce capsules containing any seed.

First, for the one illegitimate short-styled plant. In the course of three seasons 27 flowers were illegitimately / fertilized with pollen from the same plant, and they yielded only a single capsule, which, however, contained a rather large number of seeds for a union of this nature, namely, 23. As a standard of comparison I may state that during the same three seasons 44 flowers borne by legitimate short-styled plants were self-fertilized, and yielded 26 capsules; so that the fact of the 27 flowers on the illegitimate plant having produced only one capsule proves how sterile it was. To show that the conditions of life were favourable, I will add that numerous plants of this and other species of Primula all produced an abundance of capsules whilst growing close by in the same soil with the present and following plants. The sterility of the above illegitimate short-styled plant depended on both the male and female organs being in a deteriorated condition. This was manifestly the case with the pollen; for many of the anthers were shrivelled or contabescent. Nevertheless some of the anthers contained pollen, with which I succeeded in fertilizing some flowers on the illegitimate long-styled plants immediately to be described. Four flowers on this same short-styled plant were likewise *legitimately* fertilized with pollen from one of the following long-styled plants; but only one capsule was produced, containing 26 seeds; and this is a very low number for a legitimate union.

With respect to the five illegitimate long-styled plants of the first generation, derived from the above self-fertilized short-styled and long-styled parents, their fertility was observed during the same three years. These five plants, when self-fertilized, differed considerably from one another in their degree of fertility, as was the case with the illegitimate long-styled plants of *Lythrum salicaria*; and their fertility / varied much according to the season. I may premise, as a standard of comparison, that during the same years 56 flowers on legitimate long-styled plants of the same age and grown in the same soil, were fertilized with their own pollen, and yielded 27 capsules; that is, 48 per cent. On one of the five illegitimate long-styled plants 36 flowers were self-fertilized in the course of the three years, but they did not produce a single capsule. Many of the anthers on this plant were contabescent: but some seemed to contain sound pollen. Nor were the female organs quite impotent; for I obtained from a *legitimate* cross one capsule with good seed. On a second illegitimate long-styled plant 44 flowers were fertilized during the same years with their own pollen, but they produced only a single capsule. The third and fourth plants were in a very slight degree more productive. The fifth and last plant was decidedly more fertile; for 42 self-fertilized flowers yielded 11 capsules. Altogether, in the course of the three years, no less than 160 flowers on these five illegitimate long-styled plants were fertilized with their own pollen, but they yielded only 22 capsules. According to the standard above given, they ought to have yielded 80 capsules.

These 22 capsules contained on an average 15·1 seeds I believe, subject to the doubts before specified, that with legitimate plants the average number from a union of this nature would have been above 20 seeds. Twenty-four flowers on these same five illegitimate long-styled plants were legitimately fertilized with pollen from the above-described illegitimate short-styled plant, and produced only 9 capsules, which is an extremely small number for a legitimate union. These 9 capsules, however, contained an average of 38 apparently good seeds, which is as large a number as / legitimate plants sometimes yield. But this high average was almost certainly false; and I mention the case for the sake of showing the difficulty of arriving at a fair result; for this average mainly depended on two capsules containing the extraordinary numbers of 75 and 56 seeds; these seeds, however, though I felt bound to count them, were so poor that, judging from trials made in other cases, I do not suppose that one would have germinated; and therefore they ought not to have been included.

Lastly, 20 flowers were legitimately fertilized with pollen from a legitimate plant, and this increased their fertility; for they produced 10 capsules. Yet this is but a very small proportion for a legitimate union.

There can, therefore, be no doubt that these five long-styled plants and the one short-styled plant of the first illegitimate generation were extremely sterile. Their sterility was shown, as in the case of hybrids, in another way, namely, by their flowering profusely, and especially by the long endurance of the flowers. For instance, I fertilized many flowers on these plants, and fifteen days afterwards (viz. on 22 March) I fertilized numerous long-styled and short-styled flowers on common cowslips growing close by. These latter flowers, on 8 April, were withered, whilst most of the illegitimate flowers remained quite fresh for several days subsequently; so that some of these illegitimate plants, after being fertilized, remained in full bloom for above a month.

We will now turn to the fertility of the 53 illegitimate long-styled grandchildren, descended from the long-styled plant which was first fertilized with its own pollen. The pollen in two of these plants included a multitude of small and shrivelled grains. Nevertheless they were not very sterile; for 25 flowers, fertilized / with their own pollen, produced 15 capsules, containing an average of 16·3 seeds. As already stated, the probable average with legitimate plants for a union of this nature is rather above 20 seeds. These plants were remarkably healthy and vigorous, as long as they were kept under highly favourable conditions in pots in the greenhouse; and such treatment greatly increases the fertility of the cowslip. When these same plants were planted during the next year (which, however, was an unfavourable one), out of doors in good soil, 20 self-fertilized flowers produced only 5 capsules, containing extremely few and wretched seeds.

Four long-styled great-grandchildren were raised from the self-fertilized grandchildren, and were kept under the same highly favourable conditions in the greenhouse; 10 of their flowers were fertilized with own-form pollen and yielded the large proportion of 6 capsules, containing on an average 18·7 seeds. From these seeds 20 long-styled great-great-grandchildren were raised, which were likewise kept in the greenhouse. Thirty of their flowers were fertilized with their own pollen and yielded 17 capsules, containing on an average no less than 32, mostly fine seeds. It appears, therefore, that the fertility of these plants of the fourth illegitimate generation, as long as they were kept under highly favourable conditions, had not decreased, but had rather increased. The result, however, was widely different when they were planted out of doors in good soil, where other cowslips grew vigorously and were completely fertile; for these illegitimate plants now became much dwarfed in stature and extremely sterile, notwithstanding that they were exposed to the visits of insects, and must have been legitimately fertilized by the surrounding legitimate plants. A whole / row of these plants of the fourth illegitimate generation, thus freely exposed and legitimately fertilized, produced only 3 capsules, containing on an average only 17 seeds. During the ensuing winter almost all these plants died, and the few survivors were miserably unhealthy, whilst the surrounding legitimate plants were not in the least injured.

The seeds from the great-great-grandchildren were sown, and 8 long-styled and 2 short-styled plants of the fifth illegitimate generation raised. These whilst still in the greenhouse produced smaller leaves and shorter flower-stalks than some legitimate plants with which they grew in competition; but it should be observed that the latter were the product of a cross with a fresh stock – a circumstance which by itself would have added much to their vigour.[11] When these illegitimate plants were transferred to fairly good soil out of doors, they became during the two following years much more dwarfed in stature and produced very few flower-stems; and although they must have been legitimately fertilized by insects, they yielded capsules, compared with those produced by the surrounding legitimate plants, in the ratio only of 5 to 100! It is therefore certain that illegitimate fertilization, continued during successive generations, affects the powers of growth and fertility of *P. veris* to an extraordinary degree; more especially when the plants are exposed to ordinary conditions of life, instead of being protected in a greenhouse.

Equal-styled red variety of P. veris

Mr Scott has described[12] a plant of this kind growing in the Botanic Garden of Edinburgh. He states that it was highly self-fertile, although insects / were excluded; and he explains this fact by showing, first, that the anthers and stigma are in close apposition, and that the stamens in length, position and size of their pollen grains resemble those of the short-styled form, whilst the pistil resembles that of the long-styled form, both in length and in the structure of the stigma. Hence the self-union of this variety is, in fact, a legitimate union, and consequently is highly fertile. Mr Scott further states that this variety

[11] For full details of this experiment, see my *Effects of Cross and Self-fertilisation*, 1876, p. 220.
[12] *Proc. Linn. Soc.*, vol. viii (1864), p. 105.

yielded very few seeds when fertilized by either the long- or short-styled common cowslip, and, again, that both forms of the latter, when fertilized by the equal-styled variety, likewise produced very few seeds. But his experiments with the cowslip were few, and my results do not confirm his in any uniform manner.

I raised twenty plants from self-fertilized seed sent me by Mr Scott; and they all produced red flowers, varying slightly in tint. Of these, two were strictly long-styled both in structure and in function; for their reproductive powers were tested by crosses with both forms of the common cowslip. Six plants were equal-styled; but on the same plant the pistil varied a good deal in length during different seasons. This was likewise the case, according to Mr Scott, with the parent plant. Lastly, twelve plants were in appearance short-styled; but they varied much more in the length of their pistils than ordinary short-styled cowslips, and they differed widely from the latter in their powers of reproduction. Their pistils had become short-styled in structure, whilst remaining long-styled in function. Short-styled cowslips, when insects are excluded, are extremely barren: for instance, on one occasion six fine plants produced only about 50 seeds (that is, less than the product of two good capsules), and on another occasion not a single capsule. Now, when the above twelve apparently short-styled seedlings were similarly treated, nearly all produced a great abundance of capsules, containing numerous seeds, which germinated remarkably well. Moreover three of these plants, which during the first year were furnished with quite short pistils, on the following year produced pistils of extraordinary length. The greater number, therefore, of these short-styled plants could not be distinguished in function from the equal-styled variety. The anthers in the six equal-styled and in the apparently twelve short-styled plants were seated high up in the corolla, as in the true short-styled cowslip; and the pollen grains resembled those of the same form in their large size, but were mingled / with a few shrivelled grains. In function this pollen was identical with that of the short-styled cowslip; for ten long-styled flowers of the common cowslip, legitimately fertilized with pollen from a true equal-styled variety, produced six capsules, containing on an average 34·4 seeds; whilst seven capsules on a short-styled cowslip illegitimately fertilized with pollen from the equal-styled variety, yielded an average of only 14·5 seeds.

As the equal-styled plants differ from one another in their powers of reproduction, and as this is an important subject, I will give a few details with respect to five of them. First, an equal-styled plant, protected from insects (as was done in all the following cases, with one stated exception), spontaneously produced numerous capsules, five of which gave an average of 44·8 seeds, with a maximum in one capsule of 57. But six capsules, the product of fertilization with pollen from a short-styled cowslip (and this is a legitimate union), gave an average of 28·5 seeds, with a maximum of 49; and this is a much lower average than might have been expected. Secondly, nine capsules from another equal-styled plant, which had not been protected from insects, but probably was self-fertilized, gave an average of 45·2 seeds, with a maximum of 58. Thirdly, another plant which had a very short pistil in 1865, produced spontaneously many capsules, six of which contained an average of 33·9 seeds, with a maximum of 38. In 1866 this same plant had a pistil of wonderful length; for it projected quite above the anthers, and the stigma resembled that of the long-styled form. In this condition it produced spontaneously a vast number of fine capsules, six of which contained almost exactly the same average number as before, viz. 34·3, with a maximum of 38. Four flowers on this plant, legitimately fertilized with pollen from a short-styled cowslip, yielded capsules with an average of 30·2 seeds. Fourthly, another short-styled plant spontaneously produced in 1865 an abundance of capsules, ten of which contained an average of 35·6 seeds, with a maximum of 54. In 1866 this same plant had become in all respects long-styled, and ten capsules gave almost exactly the same average as before, viz. 35·1 seeds, with a maximum of 47. Eight flowers on this plant, legitimately fertilized with pollen from a short-styled cowslip, produced six capsules, with the high average of 53 seeds, and the high maximum of 67. Eight flowers were also fertilized with pollen from a long-styled cowslip / (this being an illegitimate union), and produced seven capsules containing an average of 24·4 seeds, with a maximum of 32. The fifth and last plant remained in the same condition during both years: it had a pistil rather longer than that of the true short-styled form, with the stigma smooth, as it ought to be in this form, but abnormal in shape, like a much-elongated inverted cone. It produced spontaneously many capsules, five of which, in 1865, gave an average of only 15·6 seeds; and in 1866 ten capsules still gave an average only a little higher, viz. of 22·1, with a maximum of 30. Sixteen flowers were fertilized with pollen from a long-styled cowslip, and produced 12 capsules, with an average of 24·9 seeds, and a maximum of 42. Eight flowers were fertilized with pollen from a short-styled cowslip, but yielded only two capsules, containing 18 and 23 seeds. Hence this plant, in function and partially in structure, was in an almost exactly intermediate state between the long-styled and short-styled form, but inclining towards the short-styled; and this accounts for the low average of seeds which it produced when spontaneously self-fertilized.

The foregoing five plants thus differ much from one another in the nature of their fertility. In two individuals a great difference in the length of the pistil during two succeeding years made no difference in the number of seeds produced. As all five plants possessed the male organs of the short-styled form in a perfect state, and the female organs of the long-styled form in a more or less complete state, they spontaneously produced a surprising number of capsules, which generally contained a large average of remarkably fine seeds. With ordinary cowslips, *legitimately fertilized*, I once obtained from plants cultivated in the greenhouse the high average, from seven capsules, of 58·7 seeds, with a maximum in one capsule of 87 seeds; but from plants grown out of doors I never obtained a higher average than 41 seeds. Now two of the equal-styled plants, grown out of doors and spontaneously *self-fertilized*, gave averages of 44 and 45 seeds; but this high fertility may perhaps be in part attributed to the stigma receiving pollen from the surrounding anthers at exactly the right period. Two of these plants, fertilized with pollen from a short-styled cowslip (and this in fact is a legitimate union), gave a lower average than when self-fertilized. On the other hand, another plant, when similarly fertilized by a cowslip, yielded the unusually high average of 53 seeds, with a maximum of 67. Lastly, as we have just seen, one of these plants was in / an almost exactly intermediate condition in its female organs between the long-

and short-styled forms, and consequently, when self-fertilized, yielded a low average of seed. If we add together all the experiments which I made on the equal-styled plants, 41 spontaneously self-fertilized capsules (insects having been excluded) gave an average of 34 seeds, which is exactly the same number as the parent plant yiel'ed in Edinburgh. Thirty-four flowers, fertilized with pollen from the short-r.yled cowslip (and this is an analogous union), produced 17 capsules, containing an average of 33·8 seeds. It is a rather singular circumstance, for which I cannot account, that 20 flowers, artificially fertilized on one occasion with pollen from the same plants, yielded only ten capsules, containing the low average of 26·7 seeds.

As bearing on inheritance, it may be added that 72 seedlings were raised from one of the red-flowered, strictly equal-styled, self-fertilized plants descended from the similarly characterized Edinburgh plant. These 72 plants were therefore grandchildren of the Edinburgh plant, and they all bore, as in the first generation, red flowers, with the exception of one plant, which reverted in colour to the common cowslip. In regard to structure, nine plants were truly long-styled and had their stamens seated low down in the corolla in the proper position; the remaining 63 plants were equal-styled, though the stigma in about a dozen of them stood a little below the anthers. We thus see that the anomalous combination in the same flower, of the male and female sexual organs which properly exist in the two distinct forms, was inherited with much force. Thirty-six seedlings were also raised from long- and short-styled common cowslips, crossed with pollen from the equal-styled variety. Of these plants one alone was equal-styled, 20 were short-styled, but with the pistil in three of them rather too long, and the remaining 15 were long-styled. In this case we have an illustration of the difference between simple inheritance and prepotency of transmission; for the equal-styled variety, when self-fertilized, transmits its character, as we have just seen, with much force, but when crossed with the common cowslip cannot withstand the greater power of transmission of the latter.

PULMONARIA

I have little to say on this genus. I obtained seeds of *P. officinalis* from a garden where the long-styled form alone grew, / and raised 11 seedlings, which were all long-styled. These plants were named for me by Dr Hooker. They differed, as has been shown, from the plants belonging to this species which in Germany were experimented on by Hildebrand;[13] for he found that the long-styled form was absolutely sterile with its own pollen, whilst my long-styled seedlings and the parent plants yielded a fair supply of seed when self-fertilized. Plants of the long-styled form of *Pulmonaria angustifolia* were, like Hildebrand's plants, absolutely sterile with their own pollen, so that I could never procure a single seed. On the other hand, the short-styled plants of this species, differently from those of *P. officinalis*, were fertile with their own pollen in a quite remarkable degree for a heterostyled plant. From seeds carefully self-fertilized I raised 18 plants, of which 13 proved short-styled and 5 long-styled.

[13] *Bot. Zeitung*, 1865, p. 13.

From flowers on long-styled plants fertilized illegitimately with pollen from the same plant, 49 seedlings were raised, and these consisted of 45 long-styled and 4 short-styled. From flowers on short-styled plants illegitimately fertilized with pollen from the same plant 33 seedlings were raised, and these consisted of 20 short-styled and 13 long-styled. So that the usual rule of illegitimately fertilized long-styled plants tending much more strongly than short-styled plants to reproduce their own form here holds good. The illegitimate plants derived from both forms flowered later than the legitimate, and were to the latter in height as 69 to 100. But as these illegitimate plants were descended from parents fertilized with their own pollen, whilst the legitimate plants were descended from parents crossed with pollen from a distinct individual, it is impossible to know how much of their difference in height and period of flowering is due to the illegitimate birth of the one set, and how much to the other set being the product of a cross between distinct plants.

Concluding remarks on the illegitimate offspring of heterostyled trimorphic and dimorphic plants

It is remarkable how closely and in how many points illegitimate unions between the two or three forms of the / same heterostyled species, together with their illegitimate offspring, resemble hybrid unions between distinct species together with their hybrid offspring. In both cases we meet with every degree of sterility, from very slightly lessened fertility to absolute barrenness, when not even a single seed-capsule is produced. In both cases the facility of effecting the first union is much influenced by the conditions to which the plants are exposed.[14] Both with hybrids and illegitimate plants the innate degree of sterility is highly variable in plants raised from the same mother-plant. In both cases the male organs are more plainly affected than the female; and we often find contabescent anthers enclosing shrivelled and utterly powerless pollen grains. The more sterile hybrids, as Max Wichura has well shown,[15] are sometimes much dwarfed in stature, and have so weak a constitution that they are liable to premature death; and we have seen exactly parallel cases with the illegitimate seedlings of Lythrum and Primula. Many hybrids are the most per-sistent and profuse flowerers, as are some illegitimate plants. When

[14] This has been remarked by many experimentalists in effecting crosses be-tween distinct species; and in regard to illegitimate unions I have given in the first chapter a striking illustration in the case of *Primula veris*.
[15] *Die Bastardbefruchtung im Pflanzenreich*, 1865.

a hybrid is crossed by either pure parent form, it is notoriously much more fertile than when crossed *inter se* or by another hybrid; so when an illegitimate plant is fertilized by a legitimate plant, it is more fertile than when fertilized *inter se* or by another illegitimate plant. When two species are crossed and they produced numerous seeds, we expect as a general rule that their hybrid offspring will be moderately fertile; but if the parent species produced extremely few seeds, we expect that the hybrids will be very / sterile. But there are marked exceptions, as shown by Gärtner, to these rules. So it is with illegitimate unions and illegitimate offspring. Thus the mid-styled form of *Lythrum salicaria*, when illegitimately fertilized with pollen from the longest stamens of the short-styled form, produced an unusual number of seeds; and their illegitimate offspring were not at all, or hardly at all, sterile. On the other hand, the illegitimate offspring from the long-styled form, fertilized with pollen from the shortest stamens of the same form; yielded few seeds, and the illegitimate offspring thus produced were very sterile; but they were more sterile than might have been expected relatively to the difficulty of effecting the union of the parent sexual elements. No point is more remarkable in regard to the crossing of species than their unequal reciprocity. Thus species A will fertilize B with the greatest ease; but B will not fertilize A after hundreds of trials. We have exactly the same case with illegitimate unions; for the mid-styled *Lythrum salicaria* was easily fertilized by pollen from the longest stamens of the short-styled form, and yielded many seeds; but the latter form did not yield a single seed when fertilized by the longest stamens of the mid-styled form.

Another important point is prepotency. Gärtner has shown that when a species is fertilized with pollen from another species, if it be afterwards fertilized with its own pollen, or with that of the same species, this is so prepotent over the foreign pollen that the effect of the latter, though placed on the stigma some time previously, is entirely destroyed. Exactly the same thing occurs with the two forms of a heterostyled species. Thus several long-styled flowers of *Primula veris* were fertilized illegitimately with pollen from another plant of the same form, and twenty-four hours / afterwards legitimately with pollen from a short-styled dark-red polyanthus which is a variety of *P. veris*; and the result was that every one of the thirty seedlings thus raised bore flowers more or less red, showing plainly how prepotent the legitimate pollen from a short-styled plant was over the illegitimate pollen from a long-styled plant.

In all the several foregoing points the parallelism is wonderfully close between the effects of illegitimate and hybrid fertilization. It is hardly an exaggeration to assert that seedlings from an illegitimately fertilized heterostyled plant are hybrids formed within the limits of one and the same species. This conclusion is important, for we thus learn that the difficulty in sexually uniting two organic forms and the sterility of their offspring, afford no sure criterion of so-called specific distinctness. If any one were to cross two varieties of the same form of Lythrum or Primula for the sake of ascertaining whether they were specifically distinct, and he found that they could be united only with some difficulty, that their offspring were extremely sterile, and that the parents and their offspring resembled in a whole series of relations crossed species and their hybrid offspring, he might maintain that his varieties had been proved to be good and true species; but he would be completely decieved. In the second place, as the forms of the same trimorphic or dimorphic heterostyled species are obviously identical in general structure, with the exception of the reproductive organs, and as they are identical in general constitution (for they live under precisely the same conditions), the sterility of their illegitimate unions and that of their illegitimate offspring, must depend exclusively on the nature of the sexual elements and on their incompatibility for uniting in a particular / manner. And as we have just seen that distinct species when crossed resemble in a whole series of relations the forms of the same species when illegitimately united, we are led to conclude that the sterility of the former must likewise depend exclusively on the incompatible nature of their sexual elements, and not on any general difference in constitution or structure. We are, indeed, led to this same conclusion by the impossibilty of detecting any differences sufficient to account for certain species crossing with the greatest ease, whilst other closely allied species cannot be crossed, or can be crossed only with extreme difficulty. We are led to this conclusion still more forcibly by considering the great difference which often exists in the facility of crossing reciprocally the same two species: for it is manifest in this case that the result must depend on the nature of the sexual elements, the male element of the one species acting freely on the female element of the other, but not so in a reversed direction. And now we see that this same conclusion is independently and strongly fortified by the consideration of the illegitimate unions of trimorphic and dimorphic heterostyled plants. In so complex and obscure a subject as hybridism it is no slight gain to arrive at a definite conclusion, namely, that we must look exclusively to functional differences in the sexual elements, as the cause of the sterility of species when first crossed and of their hybrid offspring. It was this consideration which led me to make the many obervations recorded in this chapter, and which in my opinion make them worthy of publication. /

CHAPTER VI

CONCLUDING REMARKS ON
HETEROSTYLED PLANTS

The essential character of heterostyled plants – Summary of the differences in fertility between legitimately and illegitimately fertilized plants –
Diameter of the pollen grains, size of anthers and structure of stigma in
the different forms – Affinities of the genera which include heterostyled
species – Nature of the advantages derived from heterostylism – The
means by which plants became heterostyled – Transmission of form –
Equal-styled varieties of heterostyled plants – Final remarks.

In the foregoing chapters all the heterostyled plants known to me have
been more or less fully described. Several other cases have been
indicated, especially by Professor Asa Gray and Kuhn,[1] in which the
individuals of the same species differ in the length of their stamens
and pistils; but as I have been often deceived by this character taken
alone, it seems to me the more prudent course not to rank any species
as heterostyled, unless we have evidence of more important differences between the forms, as in the diameter of the pollen grains, or in
the structure of the stigma. The individuals of many ordinary
hermaphrodite plants habitually fertilize one another, owing to their
male and female organs being mature at different periods, or to the
structure of the parts, or to self-sterility, etc.; and so it is with many
hermaphrodite animals, for instance, land-snails or earth-worms; but
in all these cases any one individual can fully fertilize or be fertilized /
by any other individual of the same species. This is not so with heterostyled plants; a long-styled, mid-styled or short-styled plant cannot
fully fertilize or be fertilized by any other individual, but only by one
belonging to another form. Thus the essential character of plants
belonging to the heterostyled class is that the individuals are divided
into two or three bodies, like the males and females of dioecious plants

---

[1] Asa Gray, *American Journ. of Science*, 1865, p. 101; and elsewhere as already
referred to. Kuhn, *Bot. Zeitung*, 1867, p. 67.

or of the higher animals, which exist in approximately equal numbers and are adapted for reciprocal fertilization. The existence, therefore, of two or three bodies of individuals, differing from one another in the above more important characteristics, offers by itself good evidence that the species is heterostyled. But absolutely conclusive evidence can be derived only from experiments, and by finding that pollen must be applied from the one form to the other in order to ensure complete fertility.

In order to show how much more fertile each form is when legitimately fertilized with pollen from the other form (or in the case of trimorphic species, with the proper pollen from one of the two other forms) than when illegitimately fertilized with its own-form pollen, I will append a Table (33) giving a summary of the results in all the cases hitherto ascertained. The fertility of the unions may be judged by two standards, namely, by the proportion of flowers which, when fertilized in the two methods, yield capsules, and by the average number of seeds per capsule. When there is a dash in the left-hand column opposite to the name of the species, the proportion of the flowers which yielded capsules was not recorded.

The two or three forms of the same heterostyled species do not differ from one another in general habit or foliage, as sometimes, though rarely, happens with / the two sexes of dioecious plants. Nor does the calyx differ, but the corolla sometimes differs slightly in shape, owing to the different position of the anthers. In Borreria the hairs within the tube of the corolla are differently situated in the long-styled and short-styled forms. In Pulmonaria there is a slight difference in the size of / the corolla, and in Pontederia in its colour. In the reproductive organs the differences are much greater and more important. In the one form the stamens may be all of the same length, and in the other graduated in length, or alternately longer and shorter. The filaments may differ in colour and thickness, and are sometimes nearly thrice as long in the one form as in the other. They adhere also for very different proportional lengths to the corolla. The anthers sometimes differ much in size in the two forms. Owing to the rotation of the filaments, the anthers, when mature, dehisce towards the circumference of the flower in one form of Faramea, and towards the centre in the other form. The pollen grains sometimes differ conspicuously in colour, and often to an extraordinary degcree in diameter. They differ also somewhat in shape, and apparently in their contents, as they are unequally opaque. In the short-styled form of

TABLE 33

Fertility of the legitimate unions taken together, compared with that of the illegitimate unions together. The fertility of the legitimate unions, as judged by both standards, is taken as 100

| | ILLEGITIMATE UNIONS | |
| | Proportional number of flowers which produced capsules | Average number of seeds per capsule |
Name of species		
Primula veris	69	65
P. elatior	27	75
P. vulgaris	60	54
P. Sinensis	84	63
P. Sinensis (second trial)	0	53
P. Sinensis (Hildebrand)	100	42
P. auricula (Scott)	80	15
P. Sikkimensis (Scott)	95	31
P. cortusoides (Scott)	74	66
P. involucrata (Scott)	72	48
P. farinosa (Scott)	71	44
Average of the nine species of Primula	88·4	69
Hottonia palustris (H. Müller)	—	61
Linum grandiflorum (the difference probably is much greater)	—	69
L. perenne	—	20
L. perenne (Hildebrand)	0	0
Pulmonaria officinalis (German stock, Hildebrand)	0	0
Pulmonaria angustifolia	35	32
Mitchella repens	20	47
Borreria, Brazilian sp.	—	0
Polygonum fagopyrum	—	46
Lythrum salicaria	33	46
Oxalis Valdiviana (Hildebrand)	2	34
O. Regnelli (Hildebrand)	0	0
O. speciosa	15	49

Faramea the pollen grains are covered with sharp points, so as to cohere readily together or to an insect; whilst the smaller grains of the long-styled form are quite smooth.

With respect to the pistil, the style may be almost thrice as long in the one form as in the other. In Oxalis it sometimes differs in hairiness in the three forms. In Linum the pistils either diverge and pass out

179

between the filaments, or stand nearly upright and parallel to them. The stigmas in the two forms often differ much in size and shape, and more especially in the length and thickness of their papillae; so that the surface may be rough or quite smooth. Owing to the rotation of the styles, the papillose surface of the stigma is turned outwards in one form of *Linum perenne*, and inwards in the other form. In flowers of the same age of *Primula veris* the ovules are larger in the long-styled than in the short-styled form. The / seeds produced by the two or three forms often differ in number, and sometimes in size and weight; thus, five seeds from the long-styled form of *Lythrum salicaria* equal in weight six from the mid-styled and seven from the short-styled form. Lastly, short-styled plants of *Pulmonaria officinalis* bear a larger number of flowers, and these set a larger proportional number of fruit, which however yield a lower average number of seed, than the long-styled plants. With heterostyled plants we thus see in how many and in what important characters the forms of the same undoubted species often differ from one another – characters which with ordinary plants would be sufficient to distinguish species of the same genus.

As the pollen grains of ordinary species belonging to the same genus generally resemble one another closely in all respects, it is worth while to show, in the following table (34), the difference in diameter between the grains from the two or three forms of the same heterostyled species in the forty-three cases in which this was ascertained. But it should be observed that some of the following measurements are only approximately accurate, as only a few grains were measured. In several cases, also, the grains had been dried and were then soaked in water. Whenever they were of an elongated shape their longer diameters were measured. The grains from the short-styled plants are invariably larger than those from the long-styled, whenever there is any difference between them. The diameter of the former is represented in the table by the number 100.

We here see that, with seven or eight exceptions out of the forty-three cases, the pollen grains from one form are larger than those from the other form of the same species. The extreme difference is as 100 to 55; / and we should bear in mind that in the case of spheres differing to this degree in diameter, their contents differ in the ratio of six to one. With all the species in which the grains differ in diameter, there is no exception to the rule that those from the / anthers of the short-styled form, the tubes of which have to penetrate the longer pistil of the long-styled form, are larger than the grains from the other

TABLE 34

*Relative diameter of the pollen grains from the forms of the same heterostyled species;
those from the short-styled form being represented by 100*

DIMORPHIC SPECIES

	From the long-styled form		From the long-styled form
Primula veris	67	Cordia (sp.?)	100
Primula vulgaris	71	*Gilia pulchella*	100
Primula Sinensis (Hildebrand)	57	*Gilia micrantha*	81
Primula auricula	71	*Sethia acuminata*	83
Hottonia palustris (H. Müller)	61	Erythroxylum (sp.?)	93
Hottonia palustris (self)	64	*Cratoxylon formosum*	86
Linum grandiflorum	100	*Mitchella repens*, pollen grains	
Linum perenne (diameter		of the long-styled a	
variable)	100(?)	little smaller	
Linum flavum	100	Borreria (sp.?)	92
Pulmonaria officinalis	78	Faramea (sp.?)	67
Pulmonaria angustifolia	91	Suteria (sp.?) (Fritz Müller)	75
Polygonum fagopyrum	82	*Houstonia coerulea*	72
Leucosmia Burnettiana	99	Oldenlandia (sp.?)	78
Aegiphila elata	62	Hedyotis (sp.?)	88
Menyanthes trifoliata	84	Coccocypselum (sp.?)	
Limnanthemum Indicum	100	(F. Müller)	100
Villarsia (sp.?)	75	Lipostoma (sp.?)	80
Forsythia suspensa	94	*Cinchona micrantha*	91

TRIMORPHIC SPECIES

Ratio expressing the extreme differences in diameter of the pollen grains from the two sets of anthers in the three forms		*Ratio between the diameters of the pollen grains of the two sets of anthers in the same form*	
		Oxalis rosea, long-styled form	
Lythrum salicaria	60	(Hildebrand)	83
Nesaea verticillata	65	*Oxalis compressa*, short-styled	
Oxalis Valdiviana (Hildebrand)	71	form	83
Oxalis Regnelli	78	Pontederia (sp.?) short-styled	
Oxalis speciosa	69	form	87
Oxalis sensitiva	84	Pontederia other sp., mid-styled	
Pontederia (sp.?)	55	form	86

form. Thus curious relation led Delpino[2] (as it formerly did me) to believe that the larger size of the grains in the short-styled flowers is connected with the greater supply of matter needed for the development of their longer tubes. But the case of Linum, in which the grains

[2] *Sull' Opera, la Distribuzione dei Sessi nelle Piante*, etc., 1867, p. 17.

of the two forms are of equal size, whilst the pistil of the one is about twice as long as that of the other, made me from the first feel very doubtful with respect to this view. My doubts have since been strengthened by the cases of Limnanthemum and Coccocypselum, in which the grains are of equal size in the two forms; whilst in the former genus the pistil is nearly thrice and in the latter twice as long as in the other form. In those species in which the grains are of unequal size in the two forms, there is no close relationship between the degree of their inequality and that of their pistils. Thus in *Pulmonaria officinalis* and in Erythroxylum the pistil in the long-styled form is about twice the length of that in the other form, whilst in the former species the pollen grains are as 100 to 78, and in the latter as 100 to 93 in diameter. In the two forms of Suteria the pistil differs but little in length, whilst the pollen grains are as 100 to 75 in diameter. These cases seem to prove that the difference in size between the grains in the two forms is not determined by the length of the pistil, down which the tubes have to grow. That which plants in general there is no close relationship between / the size of the pollen grains and the length of the pistil is manifest; for instance, I found that the distended grains of *Datura arborea* were 0·00243 of an inch in diameter, and the pistil no less than 9·25 inches in length; now the pistil in the small flowers of *Polygonum fagopyrum* is very short, yet the larger pollen grains from the short-styled plants had exactly the same diamater as those from the Datura, with its enormously elongated pistil.

Notwithstanding these several considerations, it is difficult quite to give up the belief that the pollen grains from the longer stamens of heterostyled plants have become larger in order to allow of the development of longer tubes; and the foregoing opposing facts may possibly be reconciled in the following manner. The tubes are at first developed from matter contained within the grains, for they are sometimes exserted to a considerable length, before the grains have touched the stigma; but botanists believe that they afterwards draw nourishment from the conducting tissue of the pistil. It is hardly possible to doubt that this must occur in such cases as that of the Datura, in which the tubes have to grow down the whole length of the pistil, and therefore to a length equalling 3,806 times the diameter of the grains (namely, 0·00243 of an inch) from which they are protruded. I may here remark that I have seen the pollen grains of a willow, immersed in a very weak solution of honey, protrude their tubes, in the course of twelve hours, to a length thirteen times as great as the diameter of the grains. Now if we suppose that the tubes in some heterostyled species are developed wholly or almost wholly from matter contained within the grains, while in other species from matter yielded by the pistil, we can see that in the former case it would be necessary / that the grains of the two forms should differ in size relatively to the length of the pistil which the tubes have to penetrate, but that in the latter case it would not be necessary that the grains should thus differ. Whether this explanation can be considered satisfactory must remain at present doubtful.

There is another remarkable difference between the forms of several heterostyled species, namely in the anthers of the short-styled flowers, which contain the larger pollen grains, being longer than those of the long-styled flowers. This is the case with *Hottonia palustris* in the ratio of 100 to 83. With *Limnanthemum Indicum* the ratio is as 100 to 70. With the allied Menyanthes the anthers of the short-styled form are a little and with Villarsia conspicuously larger than those of the long-styled. With *Pulmonaria angustifolia* they vary much in size, but from an average of seven measurements of each kind the ratio is as 100 to 91. In six genera of the Rubiaceae there is a similar difference, either slightly or well marked. Lastly, in the trimorphic Pontederia the ratio is 100 to 88; the anthers from the longest stamens in the short-styled form being compared with those from the shortest stamens in the long-styled form. On the other hand, there is a similar and well-marked difference in the length of the stamens in the two forms of *Forsythia suspensa* and of *Linum flavum*; but in these two cases the anthers of the short-styled flowers are shorter than those of the long-styled. The relative size of the anthers was not particularly attended to in the two forms of the other heterostyled plants, but I believe that they are generally equal, as is certainly the case with those of the common primrose and cowslip.

The pistil differs in length in the two forms of every / heterostyled plant, and although a similar difference is very general with the stamens, yet in the two forms of *Linum grandiflorum* and of Cordia they are equal. There can hardly be a doubt that the relative length of these organs is an adaptation for the safe transportal by insects of the pollen from the one form to the other. The exceptional cases in which these organs do not stand exactly on a level in the two forms may probably be explained by the manner in which the flowers are visited. With most of the species, if there is any difference in the size of the stigma in the two forms, that of the long-styled, whatever its shape may be, is larger than that of the short-styled. But here again there are some exceptions to the rule, for in the short-styled form of *Leucosmia Burnettiana* the stigmas are longer and much narrower than those of the long-styled; the ratio between the lengths of the stigmas in the two forms being 100 to 60. In the three Rubiaceous genera, Faramea, Houstonia and Oldenlandia, the stigmas of the short-styled form are likewise somewhat longer and narrower; and in the three forms of *Oxalis sensitiva* the difference is strongly marked, for if the length of the two stigmas of the long-styled pistil be taken as 100, it will be represented in the mid- and short-styled forms by the numbers 141 and 164. As in all these cases the stigmas of the short-styled pistil are seated low down within a more or less tubular corolla, it is probable that they are better fitted by being long and narrow for brushing the pollen off the inserted proboscis of an insect.

With many heterostyled plants the stigma differs in roughness in the two forms, and when this is the case there is no known exception to the rule that the papillae on the stigma of the long-styled form are longer / and often thicker than those on that of the short-styled. For instance, the papillae on the long-styled stigma of *Hottonia palustris* are more than twice the length of those in the other form. This holds good even in the case of *Houstonia coerulea*, in which the stigmas are much shorter and stouter in the long-styled than in the short-styled form, for the papillae on the former compared with those on the latter are as 100 to 58 in length. The length of the pistil in the long-styled form of *Linum grandiflorum* varies much, and the stigmatic papillae vary in a corresponding manner. From this fact I inferred at first that in all cases the difference in length between the stigmatic papillae in the two forms was one merely of correlated growth; but this can hardly be the true or general explanation, as the shorter stigmas of the long-styled form of Houstonia have the longer papillae. It is a more probable view that the papillae, which render the stigma of the long-styled form of various species rough, serve to entangle effectually the large-sized pollen grains brought by insects from the short-styled form, thus ensuring its legitimate fertilization. This view is supported by the fact that the pollen grains from the two forms of eight species in Table 34 hardly differ in diameter, and the papillae on their stigmas do not differ in length.

The species which are at present positively or almost positively known to be heterostyled belong, as shown in Table 35, to 38 genera, widely

TABLE 35

List of genera including heterostyled species

DICOTYLEDONS		DICOTYLEDONS	
Cratoxylon	Hypericineae	Mitchella	Rubiaceae
Erythroxylum	Erythroxyleae	Diodia	,,
Sethia	,,	Borreria	,,
Linum	Geraniaceae	Spermacoce	,,
Oxalis	,,	Primula	Primulaceae
Lythrum	Lythraceae	Hottonia	,,
Nesaea	,,	Androsace	,,
Cinchona	Rubiaceae	Forsythia	Oleaceae
Bouvardia	,,	Menyanthes	Gentianaceae
Manettia	,,	Limnanthemum	,,
Hedyotis	,,	Villarsia	,,
Oldenlandia	,,	Gilia	Polemoniaceae
Houstonia	,,	Cordia	Cordieae
Coccocypselum	,,	Pulmonaria	Boragineae
Lipostoma	,,	Aegiphila	Verbenaceae
Knoxia	,,	Polygonum	Polygoneae
Faramea	,,	Thymelea	Thymeleae
Psychotria	,,		
Rudgea	,,	MONOCOTYLEDONS	
Suteria	,,	Pontederia	Pontederiaceae

distributed, throughout the world. These genera are included in fourteen Families, most of which are very distinct from one another, for they belong to nine of the several great Series, into which phanerogamic plants have been divided by Bentham and Hooker. /

In some of these families the heterostyled condition must have been acquired at a very remote period. Thus the three closely allied genera, Menyanthes, Limnanthemum, and Villarsia, inhabit respectively Europe, India, and South America. Heterostyled species of Hedyotis are found in the temperate regions of North and the tropical regions of South America. Trimorphic species of Oxalis live on both sides of the Cordillera in South America and at the Cape of Good Hope. In these and some other cases it is not probable that each species acquired its heterostyled structure independently of its close allies. If they did not do so, the three closely connected genera of the Menyantheae and the several trimorphic species of Oxalis must have inherited their structure from a common progenitor. But an immense lapse of time will have been necessary in all such cases for the modified descendants of a common progenitor to have / spread from a single centre to such

185

widely remote and separated areas. The family of the Rubiaceae contains not far short of as many heterostyled genera as all the other thirteen families together; and hereafter no doubt other Rubiaceous genera will be found to be heterostyled, although a large majority are homostyled. Several closely allied genera in this family probably owe their heterostyled structure to descent in common; but as the genera thus characterized are distributed in no less than eight of the tribes into which this family has been divided by Bentham and Hooker, it is almost certain that several of them must have become heterostyled independently of one another. What there is in the constitution or structure of the members of this family which favours their becoming heterostyled, I cannot conjecture. Some families of considerable size, such as the Boragineae and Verbenaceae, include, as far as is at present known, only a single heterostyled genus. Polygonum also is the sole heterostyled genus in its family; and though it is a very large genus, no other species except *P. fagopyrum* is thus characterized. We may suspect that it has become heterostyled within a comparatively recent period, as it seems to be less strongly so in function than the species in any other genus, for both forms are capable of yielding a considerable number of spontaneously self-fertilized seeds. Polygonum in possessing only a single heterostyled species is an extreme case; but every other genus of considerable size which includes some such species likewise contains homostyled species. Lythrum includes trimorphic, dimorphic, and homostyled species.

Trees, bushes, and herbaceous plants, both large and small, bearing single flowers or flowers in dense spikes or heads, have been rendered heterostyled. / So have plants which inhabit alpine and lowland sites, dry land, marshes and water.[3]

---

[3] Out of the 38 genera known to include heterostyled species, about eight, or 21 per cent, are more or less aquatic in their habits. I was at first struck with this fact, for I was then aware how large a proportion of ordinary plants inhabit such stations. Heterostyled plants may be said in one sense to have their sexes separated, as the former must mutually fertilize one another. Therefore it seemed worthwhile to ascertain what proportion of the genera in the Linnean classes Monoecia, Diocia and Polygamia, contained species which live 'in water, marshes, bogs or watery places'. In Sir W. J. Hooker's *British Flora* (4th edit., 1838) these three Linnean classes include 40 genera, 17 of which (i.e. 43 per cent) contain species inhabiting the just-specified stations. So that 43 per cent of those British plants which have their sexes separated are more or less aquatic in their habits, whereas only 21 per cent of heterostyled plants have such habits. I may add that the hermaphrodite classes, from Monandria to Gynandria inclusive, contain 447 genera, of which 113 are aquatic in the above sense, or only 25 per cent. It thus appears, as far as can be judged from such imperfect data that there is some connection between the separation of the sexes in plants and the watery nature of the sites which they inhabit; but that this does not hold good with heterostyled species.

When I first began to experimentize on heterostyled plants it was under the impression that they were tending to become dioecious; but I was soon forced to relinquish this notion, as the long-styled plants of Primula which, from possessing a longer pistil, larger stigma, shorter stamens with smaller pollen grains, seemed to be the more feminine of the two forms, yielded fewer seeds than the short-styled plants which appeared to be in the above respects the more masculine of the two. Moreover, trimorphic plants evidently come under the same category with dimorphic, and the former cannot be looked at as tending to become dioecious. With *Lythrum salicaria*, however, we have the curious and unique case of the mid-styled form being more feminine or less masculine in nature than the other two forms. This is shown by the large / number of seeds which it yields in whatever manner it may be fertilized, and by its pollen (the grains of which are of smaller size than those from the corresponding stamens in the other two forms) when applied to the stigma of any form producing fewer seeds than the normal number. If we suppose the process of deterioration of the male organs in the mid-styled form to continue, the final result would be the production of a female plant; and *Lythrum salicaria* would then consist of two heterostyled hermaphrodites and a female. No such case is known to exist, but it is a possible one, as hermaphrodite and female forms of the same species are by no means rare. Although there is no reason to believe that heterostyled plants are regularly becoming dioecious, yet they offer singular facilities, as will hereafter be shown, for such conversion; and this appears occasionally to have been effected.

We may feel sure that plants have been rendered heterostyled to ensure cross-fertilization, for we now know that a cross between the distinct individuals of the same species is highly important for the vigour and fertility of the offspring. The same end is gained by dichogamy or the maturation of the reproductive elements of the same flower at different periods – by dioeciousness – self-fertility – the prepotency of pollen from another individual over a plant's own pollen – and lastly, by the structure of the flower in relation to the visits of insects. The wonderful diversity of the means for gaining the same end in this case, and in many others, depends on the nature of all the previous changes through which the species had passed, and on the more or less complete inheritance of the successive adaptations of each part to the surrounding conditions. / Plants which are already well adapted by the structure of their flowers for cross-fertilization by the aid of insects often possess an irregular corolla, which has been modelled in relation to their visits; and it would have been of little or no use to such plants to have become heterostyled. We can thus understand why it is that not a single species is heterostyled in such great families as the Leguminosae, Labiatae, Scrophulariaceae, Orchideae, etc., all of which have irregular flowers. Every known heterostyled plant, however, depends on insects for its fertilization, and not on the wind; so that it is a rather surprising fact that only one genus, Pontederia, has a plainly irregular corolla.

Why some species are adapted for cross-fertilization, whilst others within the same genus are not so, or if they once were, have since lost such adaptation and in consequence are now usually self-fertilized, I have endeavoured elsewhere to explain to a certain limited extent.[4] If it be further asked why some species have been adapted for this end by being made heterostyled, rather than by any of the above specified means, the answer probably lies in the manner in which heterostylism originated – a subject immediately to be discussed. Heterostyled species, however, have an advantage over dichogamous species, as all the flowers on the same heterostyled plant belong to the same form, so that when fertilized legitimately by insects two distinct individuals are sure to intercross. On the other hand, with dichogamous plants, early or late flowers on the same individual may intercross; and a cross of this kind does hardly any or no good. Whenever it is profitable to a species to produce a / large number of seeds, and this obviously is a very common case, heterostyled will have an advantage over dioecious plants, as all the individuals of the former, whilst only half of the latter, that is the females, yield seeds. On the other hand, heterostyled plants seem to have no advantage, as far as cross-fertilization is concerned, over those which are sterile with their own pollen. They lie indeed under a slight disadvantage, for if two self-sterile plants grow near together and far removed from all other plants of the same species, they will mutually and perfectly fertilize one another, whilst this will not be the case with heterostyled dimorphic plants, unless they chance to belong to opposite forms.

It may be added that species which are trimorphic have one slight advantage over the dimorphic; for if only two individuals of a dimorphic species happen to grow near together in an isolated spot, the chances are even that both will belong to the same form, and in this

---

[4] *The Effects of Cross and Self-fertilisation*, 1876, p. 441.

case they will not produce the full number of vigorous and fertile seed-lings; all these, moreover, will tend strongly to belong to the same form as their parents. On the other hand, if two plants of the same trimorphic species happen to grow in an isolated spot, the chances are two to one in favour of their not belonging to the same form; and in this case they will legitimately fertilize one another, and yield the full complement of vigorous offspring.

The means by which plants may have been rendered heterostyled

This is a very obscure subject, on which I can throw little light, but which is worthy of discussion. It has / been shown that heterostyled plants occur in fourteen natural families, dispersed throughout the whole vegetable kingdom, and that even within the family of the Rubiaceae they are dispersed in eight of the tribes. We may therefore conclude that this structure has been acquired by various plants independently of inheritance from a common progenitor, and that it can be acquired without any great difficulty – that is, without any very unusual combination of circumstances.

It is probable that the first step towards a species becoming heterostyled is great variability in the length of the pistil and stamens, or of the pistil alone. Such variations are not very rare: with *Amsinckia spectabilis* and *Nolana prostrata* these organs differ so much in length in different individuals that until experimenting on them, I thought both species heterostyled. The stigma of *Gesneria pendulina* sometimes protrudes far beyond, and is sometimes seated beneath the anthers; so it is with *Oxalis acetosella* and various other plants. I have also noticed an extraordinary amount of difference in the length of the pistil in cultivated varieties of *Primula veris* and *vulgaris*.

As most plants are at least occasionally cross-fertilized by the aid of insects, we may assume that this was the case with our supposed varying plant; but that it would have been beneficial to it to have been more regularly cross-fertilized. We should bear in mind how impor-tant an advantage it has been proved to be to many plants, though in different degrees and ways, to be cross-fertilized. It might well happen that our supposed species did not vary in function in the right manner, so as to become either dichogamous or completely self-sterile, or in structure so as to ensure cross-fertilization. If it had / thus varied, it would never have been rendered heterostyled, as this state would then have been superfluous. But the parent species of our several existing heterostyled plants may have been, and probably were (judging from their present constitution) in some degree self-sterile; and this would have made regular cross-fertilization still more desirable.

Now let us take a highly varying species with most or all of the anthers exserted in some individuals and in others seated low down in the corolla; with the stigma also varying in position in like manner. Insects which visited such flowers would have different parts of their bodies dusted with pollen, and it would be a mere chance whether this were left on the stigma of the next flower which was visited. If all the anthers could have been placed on the same level in all the plants, then abundant pollen would have adhered to the same part of the body of the insects which frequented the flowers, and would afterwards have been deposited without loss on the stigma, if it likewise stood on the same unvarying level in all the flowers. But as the stamens and pistils are supposed to have already varied much in length and to be still varying, it might well happen that they could be reduced much more easily through Natural Selection into two sets of different lengths in different individuals, than all to the same length and level in all the individuals. We know from innumerable instances, in which the two sexes and the young of the same species differ, that there is no difficulty in two or more sets of individuals being formed which inherit different characters. In our particular case the law of compensation or balancement (which is admitted by many botanists) would tend to cause the pistil to be reduced in those individuals / in which the stamens were greatly developed, and to be increased in length in those which had their stamens but little developed.

Now if in our varying species the longer stamens were to be nearly equalized in length in a considerable body of individuals, with the pistil more or less reduced; and in another body, the shorter stamens to be similarly equalized, with the pistil more or less increased in length, cross-fertilization would be secured with little loss of pollen; and this change would be so highly beneficial to the species, that there is no difficulty in believing that it could be effected through Natural Selection. Our plant would then make a close approach in structure to a heterostyled dimorphic species; or to a trimorphic species, if the stamens were reduced to two lengths in the same flower in correspondence with that of the pistils in the other two forms. But we have not as yet even touched on the chief difficulty in understanding how heterostyled species could have originated. A completely self-sterile plant or a dichogamous one can fertilize and be fertilized by any other individual of the same species; whereas the essential character of a heterostyled plant is that an individual of one form cannot fully fertilize or be fertilized by an individual of the same form, but only by one belonging to another form.

H. Müller has suggested[5] that ordinary or homostyled plants may have been rendered heterostyled merely through the effects of habit. Whenever pollen from one set of anthers is habitually applied to a pistil of particular length in a varying species, he believes that at last the possibility of fertilization in any other / manner will be nearly or completely lost. He was led to this view by observing that Diptera frequently carried pollen from the long-styled flowers of Hottonia to the stigma of the same form, and that this illegitimate union was not nearly so sterile as the corresponding union in other heterostyled species. But this conclusion is directly opposed by some other cases, for instance by that of *Linum grandiflorum*; for here the long-styled form is utterly barren with its own-form pollen, although from the position of the anthers this pollen is invariably applied to the stigma. It is obvious that with heterostyled dimorphic plants the two female and the two male organs differ in power; for if the same kind of pollen be placed on the stigmas of the two forms, and again if the two kinds of pollen be placed on the stigmas of the same form, the results are in each case widely different. Nor can we see how this differentiation of the two female and two male organs could have been effected merely through each kind of pollen being habitually placed on one of the two stigmas.

Another view seems at first sight probable, namely, that an incapacity to be fertilized in certain ways has been specially acquired by heterostyled plants. We may suppose that our varying species was somewhat sterile (as is often the case) with pollen from its own stamens, whether these were long or short: and that such sterility was transferred to all the individuals with pistils and stamens of the same length, so that these became incapable of intercrossing freely; but that such sterility was eliminated in the case of the individuals which differed in the length of their pistils and stamens. It is, however, incredible that so peculiar a form of mutual infertility should have been specially / acquired unless it were highly beneficial to the species; and although it may be beneficial to an individual plant to be sterile with its own pollen, cross-fertilization being thus ensured, how can it be any advantage to a plant to be sterile with half its brethren, that is, with all

[5] *Die Befruchtung der Blumen*, p. 352.

the individuals belonging to the same form? Moreover, if the sterility of the unions between plants of the same form had been a special acquirement, we might have expected that the long-styled form fertilized by the long-styled would have been sterile in the same degree as the short-styled fertilized by the short-styled; but this is hardly ever the case. On the contrary, there is sometimes the widest difference in this respect, as between the two illegitimate unions of *Pulmonaria angustifolia* and of *Hottonia palustris*.

It is a more probable view that the male and female organs in two sets of individuals have been by some means specially adapted for reciprocal action; and that the sterility between the individuals of the same set or form is an incidental and purposeless result. The meaning of the term 'incidental' may be illustrated by the greater or less difficulty in grafting or budding together two plants belonging to distinct species; for as this capacity is quite immaterial to the welfare of either, it cannot have been specially acquired, and must be the incidental result of differences in their vegetative systems. But how the sexual elements of heterostyled plants came to differ from what they were whilst the species was homostyled, and how they became co-adapted in two sets of individuals, are very obscure points. We know that in the two forms of our existing heterostyled plants the pistil always differs, and the stamens generally differ in length; so does the stigma in structure, / the anthers in size, and the pollen grains in diameter. It appears, therefore, at first sight probable that organs which differ in such important respects could act on one another only in some manner for which they had been specially adapted. The probability of this view is supported by the curious rule that the greater the difference in length between the pistils and stamens of the trimorphic species of Lythrum and Oxalis, the products of which are united for reproduction, by so much the greater is the infertility of the union. The same rule applies to the two illegitimate unions of some dimorphic species, namely, *Primula vulgaris* and *Pulmonaria angustifolia*; but it entirely fails in other cases, as with *Hottonia palustris* and *Linum grandiflorum*. We shall, however, best perceive the difficulty of understanding the nature and origin of the co-adaptation between the reproductive organs of the two forms of heterostyled plants, by considering the case of *Linum grandiflorum*: the two forms of this plant differ exclusively, as far as we can see, in the length of their pistils; in the long-styled form, the stamens equal the pistil in length, but their pollen has no more effect on it than so much inorganic dust; whilst this pollen fully fertilizes the short pistil of the other form. Now, it is scarcely credible that a mere difference in the length of the pistil can make a wide difference in its capacity for being fertilized. We can believe this the less because with some plants, for instance, *Amsinckia spectabilis*, the pistil varies greatly in length without affecting the fertility of the individuals which are intercrossed. So again I observed that the same plants of *Primula veris* and *vulgaris* differed to an extraordinary degree in the length of their pistils during successive seasons; nevertheless they yielded during these seasons exactly / the same average number of seeds when left to fertilize themselves spontaneously under a net.

We must therefore look to the appearance of inner or hidden constitutional differences between the individuals of a varying species, of such a nature that the male element of one set is enabled to act efficiently only on the female element of another set. We need not doubt about the possibility of variations in the constitution of the reproductive system of a plant, for we know that some species vary so as to be completely self-sterile or completely self-sterile, either in an apparently spontaneous manner or from slightly changed conditions of life. Gärtner also has shown[6] that the individual plants of the same species vary in their sexual powers in such a manner that one will unite with a distinct species much more readily than another. But what the nature of the inner constitutional differences may be between the sets or forms of the same varying species, or between distinct species, is quite unknown. It seems therefore probable that the species which have become heterostyled at first varied so that two or three sets of individuals were formed differing in the length of their pistils and stamens and in other co-adapted characters, and that almost simultaneously the irreproductive powers became modified in such a manner that the sexual elements in one set were adapted to act on the sexual elements of another set; and consequently that these elements in the same set or form incidentally became ill-adapted for mutual interaction, as in the case of distinct species. I have elsewhere shown[7] that the sterility of species when / first crossed and of their hybrid offspring must also be looked at as merely an

---

[6] Gärtner, *Bastarderzeugung im Pflanzenreich*, 1849, p. 165.

[7] *Origin of Species*, 6th edit., p. 247; *Variation of Animals and Plants under Domestication*, 2nd edit., vol. ii, p. 169; *The Effects of Cross and Self-fertilisation*, p. 463. It may be well here to remark that, judging from the remarkable power with which abruptly changed conditions of life act on the reproductive system of most organisms, it is probable that the close adaptation of the male to the female elements in the two forms of the same heterostyled species, or in all the individuals of the same ordinary species, could be acquired only under long-continued nearly uniform conditions of life.

incidental result, following from the special co-adaptation of the sexual elements of the same species. We can thus understand the striking parallelism, which has been shown to exist between the effects of illegitimately uniting heterostyled plants and of crossing distinct species. The great difference in the degree of sterility between the various heterostyled species when illegitimately fertilized, and between the two forms of the same species when similarly fertilized, harmonizes well with the view that the result is an incidental one which follows from changes gradually effected in their reproductive systems, in order that the sexual elements of the distinct forms should act perfectly on one another.

Transmission of the two forms by heterostyled plants

The transmission of the two forms by heterostyled plants, with respect to which many facts were given in the last chapter, may perhaps be found hereafter to throw some light on their manner of development. Hildebrand observed that seedlings from the long-styled form of *Primula Sinensis* when fertilized with pollen from the same form were mostly long-styled, and many analogous cases have since been observed by me. All the known cases are given in Tables 36 and 37. /

We see in these two tables that the offspring from a form illegitimately fertilized with pollen from another plant of the same form belong, with a few exceptions, to the same form as their parents. For instance, out of 162 seedlings from long-styled plants of *Primula veris* fertilized during five generations in this manner, 156 were long-styled and only 6 short-styled. Of 69 seedlings from *P. vulgaris* similarly raised all were long-styled. So it was with 56 seedlings from the long-styled form of the trimorphic *Lythrum salicaria*, and with numerous seedlings from the long-styled form of *Oxalis rosea*. The offspring from the short-styled forms of dimorphic plants, and from both the mid-styled and short-styled forms of trimorphic plants, fertilized with their own-form pollen, likewise tend to belong to the same form as their parents, but not in so marked a manner as in the case of the long-styled form. There are three cases in Table 37, in which a form of Lythrum was fertilized illegitimately with pollen from another form; and in two of these cases all the offspring belonged to the same two forms as their parents, whilst in the third case they belonged to all three forms.

The cases hitherto given relate to illegitimate unions, but Hildebrand,

TABLE 36

Nature of the offspring from illegitimately fertilized dimorphic plants

		Number of long-styled offspring	Number of short-styled offspring
Primula veris	Long-styled form, fertilized by own-form pollen during five successive generations, produced	156	6
"	Short-styled form, fertilized by own-form pollen, produced	5	9
Primula vulgaris	Long-styled form, fertilized by own-form pollen during two successive generations, produced	69	0
Primula auricula	Short-styled form, fertilized by own-form pollen, is said to produce during successive generations offspring in about the following proportions	25	75
Primula Sinensis	Long-styled form, fertilized by own-form pollen during two successive generations, produced	52	0
"	Long-styled form, fertilized by own-form pollen (Hildebrand), produced	14	3
"	Short-styled form, fertilized by own-form pollen, produced	1	24
Pulmonaria officinalis	Long-styled form, fertilized by own-form pollen, produced	11	0
Polygonum fagopyrum	Long-styled form, fertilized by own-form pollen, produced	45	4
"	Short-styled form, fertilized by own-form pollen, produced	13	20

Fritz Müller, and myself found that a very large proportion, or all of the offspring, from a legitimate union between any two forms of the trimorphic species of Oxalis belonged to the same two forms. A similar rule therefore holds good with unions which are fully fertile, as with those of an illegitimate nature which are more or less sterile. When some of the seedlings from a heterostyled plant belong to a different form from that of its parents, Hildebrand accounts for the fact by reversion. For instance, the long-styled parent plant of *Primula veris*, from which / the 162 illegitimate seedlings in Table 36 were derived in the course of five generations, was itself no doubt derived from the union of a long-styled and a short-styled parent; and the 6 short-styled

TABLE 37

Nature of the offspring from illegitimately fertilized trimorphic plants

		Number of long-styled offspring	Number of mid-styled offspring	Number of short-styled offspring
Lythrum salicaria	Long-styled form, fertilized by own-form pollen, produced	56	0	0
„	Short-styled form, fertilized by own-form pollen, produced	1	0	8
„	Short-styled form, fertilized by pollen from mid-styled stamens of long-styled form, produced	4	0	8
„	Mid-styled form, fertilized by own-form pollen, produced	1	3	0
„	Mid-styled form, fertilized by pollen from shortest stamens of long-styled form, produced	17	8	0
„	Mid-styled form, fertilized by pollen from longest stamens of short-styled form, produced	14	8	18
Oxalis rosea	Long-styled form, fertilized during several generations by own-form pollen, produced offspring in the ratio of	100	0	0
Oxalis hedysaroides	Mid-styled form, fertilized by own-form pollen, produced	0	17	0

seedlings may be attributed to reversion to their short-styled progenitor. But it is a surprising fact in this case, and in other similar ones, that the number of the offspring which thus reverted was not larger. The fact is rendered still more strange in the particular instance of *P. veris*, for there was no reversion until four or five generations of long-styled plants had been raised. It may be seen in both tables that the long-styled form transmits its form much more faithfully than does the short-styled, when both are fertilized with their own-form pollen; and why this should be so it is difficult to conjecture, unless it be that the aboriginal parent form of most heterostyled species possessed a pistil which exceeded its own stamens considerably in length.[8] I will

[8] It may be suspected that this was the case with Primula, judging from the length of the pistil in several allied genera (see Mr J. Scott, *Journal Linn. Soc. Bot.*, vol. viii, 1864, p. 85). Herr Breitenbach found many specimens of *Primula elatior* growing in a state of nature with some flowers on the same plant long-styled, others short-styled and others equal-styled; and the long-styled form greatly preponderated in number; there being 61 of this form to 9 of the short-styled and 15 of the equal-styled.

only add that in a state of nature any single plant of a trimorphic species no doubt produces all three forms; and this may be accounted for either by its several flowers being separately fertilized by both the other forms, as Hildebrand supposes; or by pollen from both the other forms being deposited by insects on the stigma of the same flower.

Equal-styled varieties. The tendency of the dimorphic species of Primula to produce equal-styled varieties deserves notice. Cases of this kind have / been observed, as shown in the last chapter, in no less than six species, namely, *P. veris, vulgaris, Sinensis, auricula, farinosa,* and *elatior.* In the case of *P. veris*, the stamens resemble in length, position and size of their pollen grains the stamens of the short-styled form; whilst the pistil closely resembles that of the long-styled, but as it varies much in length, one proper to the short-styled form appears to have been elongated and to have assumed at the same time the functions of a long-styled pistil. Consequently the flowers are capable of spontaneous self-fertilization of a legitimate nature and yield a full complement of seed, or even more than the number produced by ordinary flowers legitimately fertilized. With *P. Sinensis*, on the other hand, the stamens resemble in all respects the shorter ones proper to the long-styled form, whilst the pistil makes a near approach to that of the short-styled, but as it varies in length, it would appear as if a long-styled pistil had been reduced in length and modified in function. The flowers in this case as in the last are capable of spontaneous legitimate fertilization, and are rather more productive than ordinary flowers legitimately fertilized. With *P. auricula* and *farinosa* the stamens resemble those of the short-styled form in length, but those of the long-styled in the size of their pollen grains; the pistil also resembles that of the long-styled, so that although the stamens and pistil are of nearly equal length, and consequently pollen is spontaneously deposited on the stigma, yet the flowers are not legitimately fertilized and yield only a very moderate supply of seed. We thus see, first, that equal-styled varieties have originated in various ways, and, secondly, that the combination of the two forms in the same flower differs in completeness. / With *P. elatior* some of the flowers on the same plant have become equal-styled, instead of all of them as in the other species.

Mr Scott has suggested that the equal-styled varieties arise through reversion to the former homostyled condition of the genus. This view is supported by the remarkable fidelity with which the equal-styled variation is transmitted after it has once appeared. I have shown in chapter XIII of my *Variation of Animals and Plants under Domestication*, that any cause which disturbs the constitution tends to induce reversion, and it is chiefly the cultivated species of Primula which become equal-styled. Illegitimate fertilization, which is an abnormal process, is likewise an exciting cause; and with illegitimately descended long-styled plants of *P. Sinensis*, I have observed the first appearance and subsequent stages of this variation. With some other plants of *P. Sinensis* of similar parentage the flowers appeared to have reverted to their original wild condition. Again, some hybrids between *P. veris* and *vulgaris* were strictly equal-styled, and others made a near approach to this structure. All these facts support the view that this variation results, at least in part, from reversion to the original state of the genus, before the species had become heterostyled. On the other hand, some considerations indicate, as previously remarked, that the aboriginal parent form of Primula had a pistil which exceeded the stamens in length. The fertility of the equal-styled varieties has been somewhat modified, being sometimes greater and sometimes less than that of a legitimate union. Another view, however, may be taken with respect to the origin of the equal-styled varieties, and their appearance may be compared with that of hermaphrodites among / animals which properly have their sexes separated; for the two sexes are combined in a monstrous hermaphrodite in a somewhat similar manner as the two sexual forms are combined in the same flower of an equal-styled variety of a heterostyled species.

Final remarks

The existence of plants which have been rendered heterostyled is a highly remarkable phenomenon, as the two or three forms of the same undoubted species differ not only in important points of structure, but in the nature of their reproductive powers. As far as structure is concerned, the two sexes of many animals and of some plants differ to an extreme degree; and in both kingdoms the same species may consist of males, females, and hermaphrodites. Certain hermaphrodite cirripedes are aided in their reproduction by a whole cluster of what I have called complemental males, which differ wonderfully from the ordinary hermaphrodite form. With ants we have males and females, and two or three castes of sterile females or workers. With Termites there are, as Fritz Müller has shown, both winged and wingless males and females, besides the workers. But in none of these cases is there any reason to believe that the several males or several females of the same species differ in their sexual powers, except in the atrophied condition of the reproductive organs in the workers of social insects. Many hermaphrodite animals must unite for reproduction, but the necessity of such union apparently depends solely on their structure. On the other hand, with hetero-styled dimorphic species there are two females and two sets of males, and with trimorphic species three females and three sets of males, which differ essentially in their sexual / powers. We shall, perhaps, best perceive the complex and extraordinary nature of the marriage arrangements of a trimorphic plant by the following illustration. Let us suppose that the individuals of the same species of ant always lived in triple communities; and that in one of these, a large-sized female (differing also in other characters) lived with six middle-sized and six small-sized males; in the second community a middle-sized female lived with six large- and six small-sized males; and in the third, a small-sized female lived with six large- and six middle-sized males. Each of these three females, though enabled to unite with any male, would be nearly sterile with her own two sets of males, and likewise with two other sets of males of the same size with her own which lived in the other two communities; but she would be fully fertile when paired with a male of her own size. Hence the thirty-six males, distri-buted by half-dozens in the three communities, would be divided into three sets of a dozen each; and these sets, as well as the three females, would differ from one another in their reproductive powers in exactly the same manner as do the distinct species of the same genus. But it is a still more remarkable fact that young ants raised from any one of the three female ants, illegitimately fertilized by a male of a different size, would resemble in a whole series of relations the hybrid offspring from a cross between two distinct species of ants. They would be dwarfed in stature, and more or less, or even utterly barren. Naturalists are so much accustomed to behold great diver-sities of structure associated with the two sexes, that they feel no surprise at almost any amount of difference; but differences in sexual nature have been thought to be the very touchstone of specific dis-tinction. We now see that such sexual differences / – the greater or less power of fertilizing and being fertilized – may characterize the co-existing individuals of the same species, in the same manner as they characterize and have kept separate those groups of individuals, produced during the lapse of ages, which we rank and denominate as distinct species. /

CHAPTER VII

POLYGAMOUS, DIOECIOUS, AND
GYNO-DIOECIOUS PLANTS

The conversion in various ways of hermaphrodite into dioecious plants –
Heterostyled plants rendered dioecious – Rubiaceae – Verbenaceae –
Polygamous and sub-dioecious plants – Euonymus – Fragaria – The two
subforms of both sexes of Rhamnus and Epigaea – Ilex – Gyno-dioecious
plants – Thymus, difference in the fertility of the hermaphrodite and female
individuals – Satureia – Manner in which the two forms probably
originated – Scabiosa and other gyno-dioecious plants – Difference in the
size of the corolla in the forms of polygamous, dioecious, and gyno-
dioecious plants.

There are several groups of plants in which all the species are
dioecious, and these exhibit no rudiments in the one sex of the organs
proper to the other. About the origin of such plants nothing is known.
It is possible that they may be descended from ancient lowly organized
forms, which had from the first their sexes separated; so that they have
never existed as hermaphrodites. There are, however, many other
groups of species and single ones, which from being allied on all sides
to hermaphrodites, and from exhibiting in the female flowers plain
rudiments of male organs, and conversely in the male flowers rudiments of female organs, we may feel sure are descended from plants
which formerly had the two sexes combined in the same flower. It is a
curious and obscure problem how and why such hermaphrodites have
been rendered bisexual.

If in some individuals of a species the stamens alone were to abort,
females and hermaphrodites would / be left existing, of which many
instances occur; and if the female organs of the hermaphrodite were
afterwards to abort, the result would be a dioecious plant. Conversely,
if we imagine the female organs alone to abort in some individuals,
males and hermaphrodites would be left; and the hermaphrodites
might afterwards be converted into females.

In other cases, as in that of the common ash tree mentioned in the

201

Introduction, the stamens are rudimentary in some individuals, the pistils in others, others again remaining as hermaphrodites. Here the modification of the two sets of organs appears to have occurred simultaneously, as far as we can judge from their equal state of abortion. If the hermaphrodites were supplanted by the individuals having separated sexes, and if these latter were equalized in number, a strictly dioecious species would be formed.

There is much difficulty in understanding why hermaphrodite plants should ever have been rendered dioecious. There would be no such conversion, unless pollen was already carried regularly by insects or by the wind from one individual to the other; for otherwise every step towards dioeciousness would lead towards sterility. As we must assume that cross-fertilization was assured before an hermaphrodite could be changed into a dioecious plant, we may conclude that the conversion has not been effected for the sake of gaining the great benefits which follow from cross-fertilization. We can, however, see that if a species were subjected to unfavourable conditions from severe competition with other plants, or from any other cause, the production of the male and female elements and the maturation of the ovules by the same individual, might prove too great a strain on its powers, and the separation of the sexes would then be highly beneficial. / This, however, would be effected only under the contingency of a reduced number of seeds, produced by the females alone, being sufficient to keep up the stock.

There is another way of looking at the subject which partially removes a difficulty that appears at first sight insuperable, namely, that during the conversion of an hermaphrodite into a dioecious plant, the male organs must abort in some individuals and the female organs in others. Yet as all are exposed to the same conditions, it might have been expected that those which varied would tend to vary in the same manner. As a general rule only a few individuals of a species vary simultaneously in the same manner; and there is no improbability in the assumption that some few individuals might produce larger seeds than the average, better stocked with nourishment. If the production of such seeds were highly beneficial to a species, and on this head there can be little doubt,[1] the variety with the large seeds would tend to increase. But in accordance with the law of compensation we might expect that the individuals which produced such seeds would, if living

---

[1] See the facts given in *The Effects of Cross and Self-fertilisation*, p. 353.

under severe conditions, tend to produce less and less pollen, so that their anthers would be reduced in size and might ultimately become rudimentary. This view occurred to me owing to a statement by Sir J. E. Smith[2] that there are female and hermaphrodite plants of *Serratula tinctoria*, and that the seeds of the former are larger than those of the hermaphrodite form. It may also be worthwhile to recall the case of the mid-styled form of *Lythrum salicaria*, which produces a larger / number of seeds than the other forms, and has somewhat smaller pollen grains which have less fertilizing power than those of the corresponding stamens in the other two forms; but whether the larger number of seeds is the indirect cause of the diminished power of the pollen, or vice versa, I know not. As soon as the anthers in a certain number of individuals became reduced in size in the manner just suggested or from any other cause, the other individuals would have to produce a larger supply of pollen, and such increased development would tend to reduce the female organs through the law of compensation, so as ultimately to leave them in a rudimentary condition; and the species would then become dioecious.

Instead of the first change occurring in the female organs we may suppose that the male ones first varied, so that some individuals produced a larger supply of pollen. This would be beneficial under certain circumstances, such as a change in the nature of the insects which visited the flowers, or in their becoming more anemophilous, for such plants require an enormous quantity of pollen. The increased action of the male organs would tend to affect through compensation the female organs of the same flower; and the final result would be that the species would consist of males and hermaphrodites. But it is of no use considering this case and other analogous ones, for, as stated in the Introduction, the co-existence of male and hermaphrodite plants is excessively rare.

It is no valid objection to the foregoing views that changes of such a nature would be effected with extreme slowness, for we shall presently see good reason to believe that various hermaphrodite plants have become or are becoming dioecious by many and excessively small steps. In the case of polygamous / species which exist as males, females and hermaphrodites, the latter would have to be supplanted before the species could become strictly dioecious; but the extinction of the hermaphrodite form would probably not be difficult, as a complete

[2] *Trans. Linn. Soc.*, vol. xiii, p. 600.

separation of the sexes appears often to be in some way beneficial. The males and females would also have to be equalized in number, or produced in some fitting proportion for the effectual fertilization of the females.

There are, no doubt, many unknown laws which govern the suppression of the male or female organs in hermaphrodite plants, quite independently of any tendency in them to become monoecious, dioecious, or polygamous. We see this in those hermaphrodites which from the rudiments still present manifestly once possessed more stamens or pistils than they now do – even twice as many, as a whole verticil has often been suppressed. Robert Brown remarks[3] that 'the order of reduction or abortion of the stamina in any natural family may with some confidence be predicted', by observing in other members of the family, in which their number is complete, the order of the dehiscence of the anthers; for the lesser permanence of an organ is generally connected with its lesser perfection, and he judges of perfection by priority of development. He also states that whenever there is a separation of the sexes in an hermaphrodite plant, which bears flowers on a simple spike, it is the females which expand first; and this he likewise attributes to the female sex being the more perfect of the two, but why the female should be thus valued he does not explain. /

Plants under cultivation or changed conditions of life frequently become sterile; and the male organs are much oftener affected than the female, though the latter alone are sometimes affected. The sterility of the stamens is generally accompanied by a reduction in their size; and we may feel sure, from a widespread analogy, that both the male and female organs would become rudimentary in the course of many generations if they failed altogether to perform their proper functions. According to Gärtner,[4] if the anthers on a plant are contabescent (and when this occurs it is always at a very early period of growth) the female organs are sometimes precociously developed. I mention this case as it appears to be one of compensation. So again is the well-known fact, that plants which increase largely by stolons or other such means are often utterly barren, with a large proportion of their pollen grains in a worthless condition.

---

[3] *Trans. Linn. Soc.*, vol. xii, p. 98. Or *Miscellaneous Works*, vol. ii, pp. 278–81.

[4] *Beiträge zur Kenntniss*, etc., p. 117 et seq. The whole subject of the sterility of plants from various causes has been discussed in my *Variation of Animals and Plants under Domestication*, chap. xviii, 2nd edit., vol. ii, pp. 146–56.

Hildebrand has shown that with hermaphrodite plants which are strongly proterandrous, the stamens in the flowers which open first sometimes abort; and this seems to follow from their being useless, as no pistils are then ready to be fertilized. Conversely the pistils in the flowers which open last sometimes abort; as when they are ready for fertilization all the pollen has been shed. He further shows by means of a series of gradations among the Compositae,[5] that a tendency from the causes just specified to produce either male or female florets, sometimes spreads to all the florets on the same head, and sometimes / even to the whole plant; and in this latter case the species becomes dioecious. In those rare instances mentioned in the Introduction, in which some of the individuals of both monoecious and hermaphrodite plants are proterandrous, others being proterogynous, their conversion into a dioecious condition would probably be much facilitated, as they already consist of two bodies of individuals, differing to a certain extent in their reproductive functions.

Dimorphic heterostyled plants offer still more strongly marked facilities for becoming dioecious; for they likewise consist of two bodies of individuals in approximately equal numbers, and what probably is more important, both the male and female organs differ in the two forms, not only in structure but in function, in nearly the same manner as do the reproductive organs of two distinct species belonging to the same genus. Now if two species are subjected to changed conditions, though of the same nature, it is notorious that they are often affected very differently; therefore the male organs, for instance, in one form of a heterostyled plant might be affected by those unknown causes which induce abortion, differently from the homologous but functionally different organs in the other form; and so conversely with the female organs. Thus the great difficulty before alluded to is much lessened in understanding how any cause whatever could lead to the simultaneous reduction and ultimate suppression of the male organs in half the individuals of a species, and of the female organs in the other half, whilst all were subjected to exactly the same conditions of life.

That such reduction or suppression has occurred in some heterostyled plants is almost certain. The Rubiaceae contained more heterostyled genera than any / other family, and from their wide distribution we may infer that many of them became heterostyled at a remote period, so that there will have been ample time for some of the species

[5] *Ueber die Geschlechtsverhältnisse bei den Compositen*, 1869, p. 89.

to have been since rendered dioecious. Asa Gray informs me that Coprosma is dioecious, and that it is closely allied through Nertera to Mitchella, which as we know is a heterostyled dimorphic species. In the male flowers of Coprosma the stamens are exserted, and in the female flowers the stigmas; so that, judging from the affinities of the above three genera, it seems probable that an ancient short-styled form bearing long stamens with large anthers and large pollen grains (as in the case of several Rubiaceous genera) has been converted into the male Coprosma; and that an ancient long-styled form with short stamens, small anthers and small pollen grains has been converted into the female form. But according to Mr Meehan,[6] Mitchella itself is dioecious in some districts; for he says that one form has small sessile anthers without a trace of pollen, the pistil being perfect; while in another form the stamens are perfect and the pistil rudimentary. He adds that plants may be observed in the autumn bearing an abundant crop of berries, and others without a single one. Should these statements be confirmed, Mitchella will be proved to be heterostyled in one district and dioecious in another.

Asperula is likewise a Rubiaceous genus, and from the published description of the two forms of *A. scoparia*, an inhabitant of Tasmania, I did not doubt that it was heterostyled; but on examining some flowers sent me by Dr Hooker they proved to be dioecious. The male flowers have large anthers and a very small / ovarium, surmounted by a mere vestige of a stigma without any style; whilst the female flowers possess a large ovarium, the anthers being rudimentary and apparently quite destitute of pollen. Considering how many Rubiaceous genera are heterostyled, it is a reasonable suspicion that this Asperula is descended from a heterostyled progenitor; but we should be cautious on this head, for there is no improbability in a homostyled Rubiaceous plant becoming dioecious. Moreover, in an allied plant, *Galium cruciatum*, the female organs have been suppressed in most of the lower flowers, whilst the upper ones remain hermaphrodite; and here we have a modification of the sexual organs without any connection with heterostylism.

Mr Thwaites informs me that in Ceylon various Rubiaceous plants are heterostyled; but in the case of Discospermum one of the two forms is always barren, the ovary containing about two aborted ovules in each loculus; whilst in the other form each loculus contains several

[6] *Proc. Acad. of Sciences of Philadelphia*, 28 July, 1868, p. 183.

perfect ovules; so that the species appears to be strictly dioecious.

Most of the species of the South American genus Aegiphila, a member of the Verbenaceae, apparently are heterostyled; and both Fritz Müller and myself thought that this was the case with *Ae. obdurata*, so closely did its flowers resemble those of the heterostyled species. But on examining the flowers, the anthers of the long-styled form were found to be entirely destitute of pollen and less than half the size of those in the other form, the pistil being perfectly developed. On the other hand, in the short-styled form the stigmas are reduced to half their proper length, having also an abnormal appearance; whilst the stamens are perfect. This plant therefore is dioecious; and we / may, I think, conclude that a short-styled progenitor, bearing long stamens exserted beyond the corolla, has been converted into the male; and a long-styled progenitor with fully developed stigmas into the female.

From the number of bad pollen grains in the small anthers of the short stamens of the long-styled form of *Pulmonaria angustifolia*, we may suspect that this form is tending to become female; but it does not appear that the other or short-styled form is becoming more masculine. Certain appearances countenance the belief that the reproductive system of *Phlox subulata* is likewise undergoing a change of some kind.

I have now given the few cases known to me in which heterostyled plants appear with some considerable degree of probability to have been rendered dioecious. Nor ought we to expect to find many such cases, for the number of heterostyled species is by no means large, at least in Europe, where they could hardly have escaped notice. Therefore the number of dioecious species which owe their origin to the transformation of heterostyled plants is probably not so large as might have been anticipated from the facilities which they offer for such conversion.

In searching for cases like the foregoing ones, I have been led to examine some dioecious or sub-dioecious plants, which are worth describing, chiefly as they show by what fine gradations hermaphrodites may pass into polygamous or dioecious species.

Polygamous, dioecious and sub-dioecious plants

Euonymus Europoeus (Celastrineae). The spindle tree is described in all the botanical works which I have consulted as an hermaphrodite. Asa Gray speaks of the flowers of the American species as perfect, whilst / those in the allied genus Celastrus are said to be 'polygamo-dioecious'.

If a number of bushes of our spindle tree be examined, about half will be found to have stamens equal in length to the pistil, with well-developed anthers; the pistil being likewise to all appearance well-developed. The other half have a perfect pistil, with the stamens short, bearing rudimentary anthers destitute of pollen; so that these bushes are females. All the flowers on the same plant present the same structure. The female corolla is smaller than that on the polleniferous bushes. The two forms are shown in the accompanying drawings.

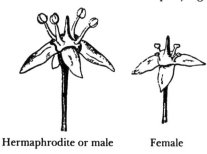

Hermaphrodite or male   Female

Fig. 12   *Euonymus Europaeus*

I did not at first doubt that this species existed under an hermaphrodite and female form; but we shall presently see that some of the bushes which appear to be hermaphrodites never produce fruit, and these are in fact males. The species, therefore, is polygamous in the sense in which I use the term, and trioicous. The flowers are frequented by many Diptera and some small Hymenoptera for the sake of the nectar secreted by the disc, but I did not see a single bee at work; nevertheless the other insects sufficed to / fertilize effectually female bushes growing at a distance or even 30 yards from any polleniferous bush.

The small anthers borne by the short stamens of the female flowers are well formed and dehisce properly, but I could never find in them a single grain of pollen. It is somewhat difficult to compare the length of the pistils in the two forms, as they vary somewhat in this respect and continue to grow after the anthers are mature. The pistils, therefore, in old flowers on a polleniferous plant are often of considerably greater length than in young flowers on a female plant. On this account the pistils from five flowers from so many hermaphrodite or male bushes were compared with those from five female bushes, before the anthers had dehisced and whilst the rudimentary ones were of a pink colour and not at all shrivelled. These two sets of pistils did not differ in length, or if there was any difference those of the polleniferous flowers were rather the longest. In one hermaphrodite plant, which produced during three years very few and poor fruit, the pistil much exceeded in length the stamens bearing perfect and as yet closed anthers; and I never saw such a case on any female plant. It is a surprising fact that the pistil in the male and in the semi-sterile hermaphrodite flowers has not been reduced in length, seeing that it performs very poorly or not at all its proper function. The stigmas in the two forms are exactly alike; and in some of the polleniferous plants which never produced any fruit I found that the surface of the stigma was viscid, so that pollen grains adhered to it and had exserted their tubes. The ovules are of equal size in the two forms. Therefore the most acute botanist, judging only by structure, would never have suspected / that some of the bushes were in function exclusively males.

Thirteen bushes growing near one another in a hedge consisted of eight females quite destitute of pollen and of five hermaphrodites with well-developed anthers. In the autumn the eight females were well covered with fruit, excepting one, which bore only a moderate number. Of the five hermaphrodites, one bore a dozen or two fruits, and the remaining four bushes several dozen; but their number was as nothing compared with those on the female bushes, for a single branch, between two and three feet in length, from one of the latter, yielded more than any one of the hermaphrodite bushes. The difference in the amount of fruit produced by the two sets of bushes is all the more striking, as from the sketches above given it is obvious that the stigmas of the polleniferous flowers can hardly fail to receive their own pollen; whilst the fertilization of the female flowers depends on pollen being brought to them by flies and the smaller Hymenoptera, which are far from being such efficient carriers as bees.

I now determined to observe more carefully during successive seasons some bushes growing in another place about a mile distant. As the female bushes were so highly productive, I marked only two of them with the letters A and B, and five polleniferous bushes with the letters C to G. I may premise that the year 1865 was highly favourable for the fruiting of all the bushes, especially for the polleniferous ones, some of which were quite barren except under such favourable conditions. The season of 1864 was unfavourable. In 1863 the female A produced 'some fruit'; in 1864 only 9; and in 1865, 97 fruit. The female B in 1863 was 'covered with fruit'; in 1864 it bore 28; and in /

1865 'innumerable very fine fruits'. I may add, that three other female trees growing close by were observed, but only during 1863, and they then bore abundantly. With respect to the polleniferous bushes, the one marked C did not bear a single fruit during the years 1863 and 1864, but during 1865 it produced no less than 92 fruit, which, however, were very poor. I selected one of the finest branches with 15 fruit, and these contained 20 seeds, or on an average 1·33 per fruit. I then took by hazard 15 fruit from an adjoining female bush, and these contained 43 seeds; that is more than twice as many, or on an average 2·86 per fruit. Many of the fruits from the female bushes included four seeds, and only one had a single seed; whereas not one fruit from the polleniferous bushes contained four seeds. Moreover, when the two lots of seeds were compared, it was manifest that those from the female bushes were the larger. The second polleniferous bush, D, bore in 1863 about two dozen fruit – in 1864 only 3 very poor fruit, each containing a single seed – and in 1865, 20 equally poor fruit. Lastly, the three polleniferous bushes, E, F, and G, did not produce a single fruit during the three years 1863, 1864, and 1865.

We thus see that the female bushes differ somewhat in their degree of fertility, and the polleniferous ones in the most marked manner. We have a perfect gradation from the female bush, B, which in 1865 was covered with 'innumerable fruits' – through the female A, which produced during the same year 97 – through the polleniferous bush C, which produced this year 92 fruits, these, however, containing a very low average number of seeds of small size – through the bush D, which produced only 20 poor fruit – to the three bushes, E, F, and G, which did not this / year, or during the two previous years, produce a single fruit. If these latter bushes and the more fertile female ones were to supplant the others, the spindle tree would be as strictly dioecious in function as any plant in the world. This case appears to me very interesting, as showing how gradually an hermaphrodite plant may be converted into a dioecious one.[7]

---

[7] According to Fritz Müller (*Bot. Zeitung*, 1870, p. 151), a Chamissoa (Amarantha-ceae) in Southern Brazil is in nearly the same state as our Euonymus. The ovules are equally developed in the two forms. In the female, the pistil is perfect, whilst the anthers are entirely destitute of pollen. In the polleniferous form, the pistil is short and the stigmas never separate from one another, so that, although their surfaces are covered with fairly well-developed papillae, they cannot be fertilized. These latter plants do not commonly yield any fruit, and are therefore in function males. Nevertheless, on one occasion Fritz Müller found flowers of this kind in which the stigmas had separated, and they produced some fruit.

Seeing how general it is for organs whicn are almost quite function-less to be reduced in size, it is remarkable that the pistils of the polleniferous plants should equal or even exceed in length those of the highly fertile female plants. This fact formerly led me to suppose that the spindle tree had once been heterostyled; the hermaphrodite and male plants having been originally long-styled, with the pistils since reduced in length, but with the stamens retaining their former dimensions; whilst the female plant had been originally short-styled, with the pistil in its present state, but with the stamens since greatly reduced and rendered rudimentary. A conversion of this kind is at least possible, although it is the reverse of that which appears actually to have occurred with some Rubiaceous genera and Aegiphila; for which these plants the short-styled form has become the male, and the long-styled the female. It is, however, a more simple view that sufficient time has not elapsed for the / reduction of the pistil in the male and hermaphrodite flowers of our Euonymus; though this view does not account for the pistils in the polleniferous flowers being sometimes longer than those in the female flowers.

Fragaria vesca, Virginiana, Chiloensis, etc (Rosaceae). A tendency to the separation of the sexes in the cultivated strawberry seems to be much more strongly marked in the United States than in Europe; and this appears to be the result of the direct action of climate on the reproductive organs. In the best account which I have seen,[8] it is stated that many of the varieties in the United States consist of three forms, namely, females, which produce a heavy crop of fruit – of hermaphrodites, which 'seldom produce other than a very scanty crop of inferior and imperfect berries' – and of males, which produce none. The most skilful cultivators plant 'seven rows of female plants, then one row of hermaphrodites, and so on throughout the field'. The males bear large, the hermaphrodites mid-sized, and the females small flowers. The latter plants produce few runners, whilst the two other forms produce many; consequently, as has been observed both in England and in the United States, the polleniferous forms increase rapidly and tend to supplant the females. We may therefore infer that much more vital force is expended in the production of ovules and fruit than in the production of pollen. Another species, the

[8] Mr Leonard Wray in *Gard. Chron.*, 1861, p. 716.

Hautbois strawberry (*F. elatior*), is more strictly dioecious; but Lindley made by selection an hermaphrodite stock.[9]

Rhamnus catharticus (Rhamneae). This plant is well / known to be dioecious. My son William found the two sexes growing in about equal numbers in the Isle of Wight, and sent me specimens, together with observations on them. Each sex consists of two subforms. The two forms of the male differ in their pistils: in some plants it is quite small, without any distinct stigma; in others the pistil is much more developed with the papillae on the stigmatic surfaces moderately large. The ovules in both kinds of males are in an aborted condition. On my mentioning this case to Professor Caspary, he examined several male plants in the botanic gardens at Königsberg, where there were no females, and sent me the accompanying drawings.

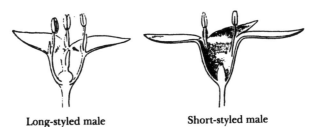

Long-styled male   Short-styled male
Fig. 13 *Rhamnus catharticus* (From Caspary)

In the English plants the petals are not so greatly reduced as represented in this drawing. My son observed that those males which had their pistils moderately well developed bore slightly larger flowers, and, what is very remarkable, their pollen grains exceeded by a little in diameter those of the males with greatly reduced pistils. This fact is opposed to the belief that the present species was once heterostyled; for in this case it might have been expected that the shorter-styled plants would have had larger pollen grains.

In the female plants the stamens are in an extremely rudimentary condition, much more so than / the pistils in the males. The pistil varies considerably in length in the female plants, so that they may be divided into two subforms according to the length of this organ. Both the petals and sepals are decidedly smaller in the females than in the

---

[9] For references and further information on this subject, see *Variation under Domestication*, chap. x, 2nd edit., vol. i, p. 375.

males; and the sepals do not turn downwards, as do those of the male flowers when mature. All the flowers on the same male or same female bush, though subject to some variability, belong to the same subform; and as my son never experienced any difficulty in deciding under which class a plant ought to be included, he believes that the two subforms of the same sex do not graduate into one another. I can form no satisfactory theory how the four forms of this plant originated.

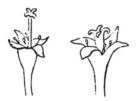

Long-styled   Short-styled
female   female

Fig. 14   *Rhamnus catharticus*

Rhamnus lanceolatus exists in the United States, as I am informed by Professor Asa Gray, under two hermaphrodite forms. In the one, which may be called the short-styled, the flowers are subsolitary, and include a pistil about two-thirds or only half as long as that in the other form; it has also shorter stigmas. The stamens are of equal length in the two forms; but the anthers of the short-styled contain rather less pollen, as far as I could judge from a few dried flowers. My / son compared the pollen grains from the two forms, and those from the long-styled flowers were to those from the short-styled, on an average from ten measurements, as 10 to 9 in diameter; so that the two hermaphrodite forms of this species resemble in this respect the two male forms of *R. catharticus*. The long-styled form is not so common as the short-styled. The latter is said by Asa Gray to be the more fruitful of the two, as might have been expected from its appearing to produce less pollen, and from the grains being of smaller size; it is therefore the more highly feminine of the two. The long-styled form produces a greater number of flowers, which are clustered together instead of being subsolitary; they yield some fruit, but as just stated are less fruitful than the other form, so that this form appears to be the more masculine of the two. On the supposition that we have here an hermaphrodite plant becoming dioecious, there are two points deserving notice; first, the greater length of the pistil in the incipient male form; and we have met with a nearly similar case in the male and hermaphrodite forms of Euonymus compared with the females. Secondly, the larger size of the pollen grains in the more masculine flowers, which perhaps may be attributed to their having retained their normal size; whilst those in the incipient female flowers have been reduced. The long-styled form of *R. lanceolatus* seems to correspond with the males of *R. catharticus* which have a longer pistil and larger pollen grains. Light will perhaps be thrown on the nature of the forms in this genus, as soon as the power of both kinds of pollen on both stigmas is ascertained. Several other species of Rhamnus are said to be dioecious[10] or sub-dioecious. / On the other hand, *R. frangula* is an ordinary hermaphrodite, for my son found a large number of bushes all bearing an equal profusion of fruit.

Epigaea repens (Ericaceae). This plant appears to be in nearly the same state as *Rhamnus catharticus*. It is described by Asa Gray[11] as existing under four forms. (1) With long style, perfect stigma, and short abortive stamens. (2) Shorter style, but with stigma equally perfect, short abortive stamens. These two female forms amounted to 20 per cent of the specimens received from one locality in Maine; but all the fruiting specimens belonged to the first form. (3) Style long, as in No. 1, but with stigma imperfect, stamens perfect. (4) Style shorter than in the last, stigma imperfect, stamens perfect. These two latter forms are evidently males. Therefore, as Asa Gray remarks, 'the flowers may be classified into two kinds, each with two modifications; the two main kinds characterized by the nature and perfection of the stigma, along with more or less abortion of the stamens; their modifications, by the length of the style'. Mr Meehan has described[12] the extreme variability of the corolla and calyx in this plant, and shows that it is dioecious. It is much to be wished that the pollen grains in the two male forms should be compared, and their fertilizing power tried on the two female forms.

Ilex aquifolium (Aquifoliaceae). In the several works which I have consulted, one author alone[13] says that the holly is dioecious. During

[10] Lecoq, *Géogr. Bot.*, vol. v, 1856, pp. 420–6.
[11] *American Journal of Science*, July, 1876. Also, *The American Naturalist*, 1876, p. 490.
[12] 'Variations in *Epigaea repens*', *Proc. Acad. Nat. Soc. of Philadelphia*, May, 1868, p. 153.
[13] Vaucher, *Hist. Phys. des Plantes d'Europe*, 1841, vol. ii, p. 11.

several years I / have examined many plants, but have never found one that was really hermaphrodite. I mention this genus because the stamens in the female flowers, although quite destitute of pollen, are, but slightly and sometimes not at all shorter than the perfect stamens in the male flowers. In the latter the ovary is small and the pistil is almost aborted. The filaments of the perfect stamens adhere for a greater length to the petals than in the female flowers. The corolla of the latter is rather smaller than that of the male. The male trees produce a greater number of flowers than the females. Asa Gray informs me that *I. opaca*, which represents in the United States our common holly, appears (judging from dried flowers) to be in a similar state; and so it is, according to Vaucher, with several other but not with all the species of the genus.

Gyno-dioecious plants

The plants hitherto described either show a tendency to become dioecious, or apparently have become so within a recent period. But the species now to be considered consist of hermaphrodites and females without males, and rarely show any tendency to be dioecious, as far as can be judged from their present condition and from the absence of species having separated sexes within the same groups. Species belonging to the present class, which I have called gyno-dioecious, are found in various widely distinct families; but are much more common in the Labiatae (as has long been noticed by botanists) than in any other group. Such cases have been noticed by myself in *Thymus serpyllum* and *vulgaris*, *Satureia hortensis*, *Origanum vulgare*, and *Mentha hirsuta*; and by others in *Nepeta glechoma*, *Mentha vulgaris* and / *aquatica*, and *Prunella vulgaris*. In these two latter species the female form, according to H. Müller, is infrequent. To these must be added *Dracocephalum Moldavicum*, *Melissa officinalis* and *clinipodium*, and *Hyssopus officinalis*.[14] In the two last-named plants the female form likewise appears to be rare, for I raised many seedlings of both, and all were

---

[14] H. Müller, *Die Befruchtung der Blumen*, 1873; and *Nature*, 1873, p. 161. Vaucher, *Plantes d'Europe*, vol. iii, p. 611. For Dracocephalum, Schimper, as quoted by Braun, *Annals and Mag. of Nat. Hist.*, 2nd series, vol. xviii, 1856, p. 380. Lecoq, *Géographie Bot. de l'Europe*, vol. viii, pp. 33, 38, 44, etc. Both Vaucher and Lecoq were mistaken in thinking that several of the plants named in the text are dioecious. They appear to have assumed that the hermaphrodite form was a male; perhaps they were deceived by the pistil not becoming fully developed and of proper length until some time after the anthers have dehisced.

hermaphrodites. It has already been remarked in the Introduction that andro-dioecious species, as they may be called, or those which consist of hermaphrodites and males, are extremely rare, or hardly exist.

Thymus serpyllum. The hermaphrodite plants present nothing particular in the state of their reproductive organs; and so it is in all the following cases. The females of the present species produce rather fewer flowers and have somewhat smaller corollas than the hermaphrodites; so that near Torquay, where this plant abounds, I could, after a little practice, distinguish the two forms whilst walking quickly past them. According to Vaucher, the smaller size of the corolla is common to the females of most or all of the above-mentioned Labiatae. The pistil of the female, though somewhat variable in length, is generally shorter, with the margins of the stigma broader and formed of more lax tissue, than that of the hermaphrodite. The stamens in the female vary excessively in length; they are generally enclosed within the tube of the / corolla, and their anthers do not contain any sound pollen; but after long search I found a single plant with the stamens moderately exerted, and their anthers contained a very few full-sized grains, together with a multitude of minute empty ones. In some females the stamens are extremely short, and their minute anthers, though divided into the two normal cells or loculi, contained not a trace of pollen: in others again the anthers did not exceed in diameter the filaments which supported them, and were not divided into two loculi. Judging from what I have myself seen and from the descriptions of others, all the plants in Britain, Germany, and near Mentone, are in the state just described; and I have never found a single flower with an aborted pistil. It is, therefore, remarkable that according to Delpino,[15] this plant near Florence is generally trimorphic, consisting of males with aborted pistils, females with aborted stamens and hermaphrodites.

I found it very difficult to judge of the proportional number of the two forms at Torquay. They often grow mingled together, but with large patches consisting of one form alone. At first I thought that the two were nearly equal in number; but on examining every plant which grew close to the edge of a little overhanging dry cliff, about 200 yards in length, I found only 12 females; all the rest, some hundreds in

---

[15] *Sull' Opera, la Distribuzione dei Sessi nelle Piante, etc.*, 1867, p. 7. With respect to Germany, H. Müller, *Die Befruchtung, etc.*, p. 327.

number, being hermaphrodites. Again, on an extensive gently sloping bank, which was so thickly covered with this plant that, viewed from the distance of half a mile it appeared of a pink colour, I could not discover a single female. Therefore the hermaphrodites / must greatly exceed in number the females, at least in the localities examined by me. A very dry station apparently favours the presence of the female form. With some of the other above-named Labiatae the nature of the soil or climate likewise seems to determine the presence of one or both forms; thus with *Nepeta glechoma*, Mr Hart found in 1873 that all the plants which he examined near Kilkenny in Ireland were females; whilst all near Bath were hermaphrodites, and near Hertford both forms were present, but with a preponderance of hermaphrodites.[16] It would, however, be a mistake to suppose that the nature of the conditions determines the form independently of inheritance; for I sowed in the same small bed seeds of *T. serpyllum*, gathered at Torquay from the female alone, and these produced an abundance of both forms. There is every reason to believe, from large patches consisting of the same form, that the same individual plant, however much it may spread, always retains the same form. In two distant gardens I found masses of the lemon-thyme (*T. citriodorus*, a var. of *T. serpyllum*), which I was informed had grown there during many years, and every flower was female.

With respect to the fertility of the two forms, I marked at Torquay a large hermaphrodite and a large female plant of nearly equal sizes, and when the seeds were ripe I gathered all the heads. The two heaps were of very nearly equal bulk; but the heads from the female plant numbered 160, and their seeds weighed 8·7 grains; whilst those from the hermaphrodite plant numbered 200, and their seeds weighed only 4·9 grains; so that the seeds from the / female plant were to those from the hermaphrodite as 100 to 56 in weight. If the relative weight of the seeds from an equal number of flower-heads from the two forms be compared, the ratio is as 100 for the female to 45 for the hermaphrodite form.

Thymus vulgaris. The common garden thyme resembles in almost every respect *T. serpyllum*. The same slight differences between the stigmas of the two forms could be perceived. In the females the stamens are not generally quite so much reduced as in the same form of *T. serpyllum*. In

[16] *Nature*, June, 1873, p. 162.

some specimens sent me from Mentone by Mr Moggridge, together with the accompanying sketches, the anthers of the female, though

Hermaphrodite   Females
Fig. 15   *Thymus vulgaris* (magnified)

small, were well formed, but they contained very little pollen, and not a single sound grain could be detected. Eighteen seedlings were raised from purchased seed, sown in the same small bed; and these consisted of seven hermaphrodites and eleven females. They were left freely exposed to the visits of bees, and no doubt every female flower was fertilized; for on placing under the microscope a large number of stigmas from female plants, / not one could be found to which pollen grains of thyme did not adhere. The seeds were carefully collected from the eleven female plants, and they weighed 98·7 grains; and those from the seven hermaphrodites 36·5 grains. This gives for an equal number of plants the ratio of 100 to 58; and we here see, as in the last case, how much more fertile the females are than the hermaphrodites. These two lots of seeds were sown separately in two adjoining beds, and the seedlings from both the hermaphrodite and female parent plants consisted of both forms.

Satureia hortensis. Eleven seedlings were raised in separate pots in a hotbed and afterwards kept in the greenhouse. They consisted of ten females and of a single hermaphrodite. Whether or not the conditions to which they had been subjected caused the great excess of females I do not know. In the females the pistil is rather longer than that of the hermaphrodite, and the stamens are mere rudiments, with minute colourless anthers destitute of pollen. The windows of the greenhouse were left open, and the flowers were incessantly visited by humble- and hive-bees. Although the ten females did not produce a single grain of pollen, yet they were all thoroughly well fertilized by the one hermaphrodite plant, and this is an interesting fact. It should be added that no other plant of this species grew in my garden. The seeds were collected from the finest female plant, and they weighed 78 grains; whilst those from the hermaphrodite, which was a rather larger plant than the female, weighed only 33·2 grains; that is, in the ratio of 100 to 43. The female form, therefore, is very much more fertile than the hermaphrodite, as in the two last cases; but the hermaphrodite was necessarily self-fertilized, and this probably diminished its fertility. /

We may now consider the probable means by which so many of the Labiatae have been separated into two forms, and the advantages thus gained. H. Müller[17] supposes that originally some individuals varied so as to produce more conspicuous flowers; and that insects habitually visited these first, and then dusted with their pollen visited and fertilized the less conspicuous flowers. The production of pollen by the latter plants would thus be rendered superfluous, and it would be advantageous to the species that their stamens should abort, so as to save useless expenditure. They would thus be converted into females. But another view may be suggested: as the production of a large supply of seeds evidently is of high importance to many plants, and as we have seen in the three foregoing cases that the females produce many more seeds than the hermaphrodites, increased fertility seems to me the more probable cause of the formation and separation of the two forms. From the data above given it follows that ten plants of *Thymus serpyllum*, if half consisted of hermaphrodites and half of females, would yield seeds compared with ten hermaphrodite plants in the ratio of 100 to 72. Under similar circumstances the ratio with *Satureia hortensis* (subject to the doubt from the self-fertilization of the hermaphrodite) would be as 100 to 60. Whether the two forms originated in certain individuals varying and producing more seed than usual, and consequently producing less pollen; or in the stamens of certain individuals tending from some unknown cause to abort, and consequently producing more seed, it is impossible to decide; but in either case, if the tendency to the increased production of seed were steadily favoured, the result would be the / complete abortion of the male organs. I shall presently discuss the cause of the smaller size of the female corolla.

Scabiosa arvensis (Dipsaceae). It has been shown by H. Müller that this species exists in Germany under an hermaphrodite and female form.[18] In my

[17] *Die Befruchtung der Blumen*, pp. 319, 326.
[18] Ibid., p. 368. The two forms occur not only in Germany, but in England and France. Lecoq (*Géographie Bot.*, 1857, vol. vi, pp. 473, 477), says that male plants as well

neighbourhood (Kent) the female plants do not nearly equal in number the hermaphrodites. The stamens of the females vary much in their degree of abortion; in some plants they are quite short and produce no pollen; in others they reach to the mouth of the corolla, but their anthers are not half the proper size, never dehisce, and contain but few pollen grains, these being colourless and of small diameter. The hermaphrodite flowers are strongly proterandrous, and H. Müller shows that, whilst all the stigmas on the same flower-head are mature at nearly the same time, the stamens dehisce one after the other; so that there is a great excess of pollen, which serves to fertilize the female plants. As the production of pollen by one set of plants is thus rendered superfluous, their male organs have become more or less completely aborted. Should it be hereafter proved that the female plants yield, as is probable more seeds than the hermaphrodites, I should be inclined to extend the same view to this plant as to the Labiatae. I have also observed the existence of two forms in our endemic *S. succisa*, and in the exotic *S. atro-purpurea*. In the latter plant, differently to what occurs in *S. arvensis*, the female flowers, especially the larger circumferential ones, are smaller than those of the hermaphrodite form. According to Lecoq, the female flower-heads of *S. succisa* are likewise smaller than those of what he calls the male plants, but which are probably hermaphrodites.

Echium vulgare (Boragineae). The ordinary hermaphrodite form appears to be proterandrous, and nothing more need be said about it. The female differs in having a much smaller corolla and shorter pistil, but a well-developed stigma. The stamens / are short; the anthers do not contain any sound pollen grains, but in their place yellow incoherent cells which do not swell in water. Some plants were in an intermediate condition; that is, had one or two or three stamens of proper length with perfect anthers, the other stamens being rudimentary. In one such plant half of one anther contained green perfect pollen grains, and the other half yellowish-green imperfect grains. Both forms produced seed, but I neglected to observe whether in equal numbers. As I thought that the state of the anthers might be due to some fungoid growth, I examined them both in the bud and mature state, but could find no trace of mycelium. In 1862 many female plants were found; and in 1864, 32 plants were collected in two localities, exactly half of which were hermaphrodites, fourteen were females, and two in an intermediate condition. In 1866, 15 plants were collected in another locality, and these consisted of four hermaphrodites and eleven females. I may add that this season was a wet one, which shows that the abortion of the stamens can hardly be due to the dryness of the sites where the plants grew, as I at one time thought probable. Seeds from an hermaphrodite were sown in my garden, and of the 23 seedlings raised, one belonged to the intermediate form, all the others being hermaphro-

---

as hermaphrodites and females co-exist; it is, however, posible that he may have been deceived by the flowers being so strongly proterandrous. From what Lecoq says, *S. succisa* likewise appears to occur under two forms in France.

dites, though two or three of them had unusually short stamens. I have consulted several botanical works, but have found no record of this plant varying in the manner here described.

Plantago lanceolata (Plantagineae). Delpino states that this plant presents in Italy three forms, which graduate from an anemophilous into an entomophilous condition. According to H. Müller,[19] there are only two forms in Germany, neither of which show any special adaptation for insect fertilization, and both appear to be hermaphrodites. But I have found in two localities in England female and hermaphrodite forms existing together; and the same fact has been noticed by others.[20] The females are less frequent than the hermaphrodites; their stamens are short, and their anthers, which are of a brighter green whilst young than those of the other form, dehisce properly, yet contain either no pollen, or a small amount of imperfect grains of variable size. All the flower-heads on a plant belong to / the same form. It is well known that this species is strongly proterogynous, and I found that the protruding stigmas of both the hermaphrodite and female flowers were penetrated by pollen tubes, whilst their own anthers were immature and had not escaped out of the bud. *Plantago media* does not present two forms; but it appears from Asa Gray's description,[21] that such is the case with four of the North American species. The corolla does not properly expand in the short-stamened form of these plants.

Cnicus Serratula Eriophorum. In the Compositae, *Cnicus Palustris* and *acaulis* are said by Sir J. E. Smith to exist as hermaphrodites and females, the former being the more frequent. With *Serratula tinctoria* a regular gradation may be followed from the hermaphrodite to the female form; in one of the latter plants the stamens were so tall that the anthers embraced the style as in the hermaphrodites, but they contained only a few grains of pollen, and these in an aborted condition; in another female, on the other hand, the anthers were much more reduced in size than is usual. Lastly, Dr Dickie has shown that with *Eriophorum angustifolium* (Cyperaceae) hermaphrodite and female forms exist in Scotland and the Arctic regions both of which yield seed.[22]

It is a curious fact that in all the foregoing polygamous, dioecious, and gyno-dioecious plants in which any difference has been observed in the size of the corolla in the two or three forms, it is rather larger in the females, which have their stamens more or less or quite rudimentary, than in the hermaphrodites or males. This holds good with Euonymus,

[19] *Die Befruchtung*, etc., p. 342.
[20] Mr C. W. Crocker in *The Gardener's Chronicle*, 1864, p. 294. Mr W. Marshall writes to me to the same effect from Ely.
[21] *Manual of the Botany of the N. United States*, 2nd edit., 1856, p. 269. See also *Americn Journal of Science*, November, 1862, p. 419, and *Proc. American Academy of Science*, 14 October, 1862, p. 53.
[22] Sir J. E. Smith, *Trans. Linn. Soc.*, vol. xiii, p. 599. Dr Dickie, *Journal Linn. Soc. Bot.* vol. ix, 1865, p. 161.

Rhamnus catharticus. Ilex, Fragaria, all or at least most of the before-named Labiatae, *Scabiosa atro-purpurea,* and *Echium vulgare.* So it is, according to Von Mohl, with *Cardamine / amara, Germanium sylvaticum, Myosotis,* and *Salvia.* On the other hand, as Von Mohl remarks, when a plant produces hermaphrodite flowers and others which are males owing to the more or less complete abortion of the female organs, the corollas of the males are not at all increased in size, or only exceptionally and in a slight degree, as in Acer.[23] It seems therefore probable that the decreased size of the female corollas in the foregoing cases is due to a tendency to abortion spreading from the stamens to the petals. We see how intimately these organs are related in double flowers, in which the stamens are readily converted into petals. Indeed some botanists believe that petals do not consist of leaves directly metamorphosed, but of metamorphosed stamens. That the lessened size of the corolla in the above case is in some manner an indirect result of the modification of the reproductive organs is supported by the fact that in *Rhamnus catharticus* not only the petals but the green and inconspicuous sepals of the female have been reduced in size; and in the strawberry the flowers are largest in the males, mid-sized in the hermaphrodites, and smallest in the females. These latter cases – the variability in the size of the corolla in some of the above species, for instance in the common thyme – together with the fact that it never differs greatly in size in the two forms – make me doubt much whether Natural Selection has come into play; that is whether, in accordance with H. Müller's belief, the advantage derived from the polleniferous flowers being visited first by insects has been sufficient to lead to a gradual reduction of the corolla of the female. We should bear in mind that as the hermaphrodite is the normal form, its corolla has / probably retained its original size.[24] An objection to the above view should not be passed over; namely, that the abortion of the stamens in the females ought to have added through the law of compensation to the size of the corolla; and this perhaps would have occurred, had not the expenditure saved by the abortion of the stamens been directed to the female reproductive organs, so as to give to this form increased fertility. /

[23] *Bot. Zeitung,* 1863, p. 326.

[24] It does not appear to me that Kerner's view (*Die Schutzmittel des Pollens,* 1873, p. 56) can be accepted in the present cases, namely that the larger corolla in the hermaphrodites and males serves to protect their pollen from rain. In the genus Thymus, for instance, the aborted anthers of the female are much better protected than the perfect ones of the hermaphrodite.

CHAPTER VIII

CLEISTOGAMIC FLOWERS

General character of cleistogamic flowers – List of the genera producing such flowers, and their distribution in the vegetable series – *Viola*, description of the cleistogamic flowers in the several species; their fertility compared with that of the perfect flowers – *Oxalis acetosella* – *O. sensitiva*, three forms of cleistogamic flowers – *Vandellia* – *Ononis* – *Impatiens* – *Drosera* – Miscellaneous observations on various other cleistogamic plants – Anemophilous species producing cleistogamic flowers – *Leersia*, perfect flowers rarely developed – Summary and concluding remarks on the origin of cleistogamic flowers – The chief conclusions which may be drawn from the observations in this volume.

It was known even before the time of Linnaeus that certain plants produced two kinds of flowers, ordinary open, and minute closed ones; and this fact formerly gave rise to warm controversies about the sexuality of plants. These closed flowers have been appropriately named cleistogamic by Dr Kuhn.[1] They are remarkable from their small size and from never opening, so that they resemble buds; their petals are rudimentary or quite aborted; their stamens are often reduced in number, with the anthers of very small size, containing few pollen grains, which have remarkably thin transparent coats, and generally emit their tubes whilst still enclosed within the anther-cells; and, lastly, the pistil is much reduced in size, with the stigma in some cases hardly at all developed. These flowers do not secrete nectar or emit any odour; from their small size, as well as from the corolla being rudimentary, they are singularly inconspicuous. Consequently / insects do not visit them; nor if they did, could they find an entrance. Such flowers are therefore invariably self-fertilized; yet they produce an abundance of seed. In several cases the young capsules bury themselves beneath the ground, and the seeds are there matured. These flowers are developed before, or after, or simultaneously with the perfect ones. Their development seems to be largely governed by the

[1] *Bot. Zeitung*, 1867, p. 65.

conditions to which the plants are exposed, for during certain seasons or in certain localities only cleistogamic or only perfect flowers are produced.

Dr Kuhn, in the article above referred to, gives a list of 44 genera including species which bear flowers of this kind. To this list I have added some genera, and the authorities are appended in a footnote. I have omitted three names, from reasons likewise given in the footnote. But it is by no means easy to decide in all cases whether certain flowers ought to be ranked as cleistogamic. For instance, Mr Bentham informs me that in the South of France some of the flowers on the vine do not fully open and yet set fruit; and I hear from two experienced gardeners that this is the case with the vine in our hot-houses; but as the flowers do not appear to be completely closed it would be imprudent to consider them as cleistogamic. The flowers of some aquatic and marsh plants, for instance of *Ranunculus aquatilis, Alisma natans,* Subularia, Illecebrum, Menyanthes, and Euryale,[2] remain closely shut as long as they are submerged, and in this condition fertilize themselves. / They behave in this manner, apparently as a protection to their pollen, and produce open flowers when exposed to the air; so that these cases seem rather different from those of true cleistogamic flowers, and have not been included in the list. Again, the flowers of some plants which are produced very early or very late in the season do not properly expand; and these might perhaps be considered as incipiently cleistogamic; but as they do not present any of the remarkable peculiarities proper to the class, and as I have not found any full record of such cases, they are not entered in the list. When, however, it is believed on fairly good evidence that the flowers on a plant in its native country do not open at any hour of the day or night, and yet set seeds capable of germination, these may fairly be considered as cleistogamic, notwithstanding that they present no peculiarities of structure. I will now give as complete a list (Table 38) of the genera containing cleistogamic species as I have been able to collect.

The first point that strikes us in considering this list of 55 genera, is that they are very widely distributed in the vegetable series. They are

---

[2] Delpino, *Sull' Opera, la Distribuzione dei Sessi nelle Piante,* etc., 1867, p. 30. Subularia, however, sometimes has its flowers fully expanded beneath the water, see Sir J. E. Smith, *English Flora,* vol. iii, 1825, p. 157. For the behaviour of Menyanthes in Russia see Gillibert in *Act. Acad. St Petersb.,* 1777, part ii, p. 45. On Euryale, *Gardener's Chronicle,* 1877, p. 280.

TABLE 38

List of genera including cleistogamic species (chiefly after Kuhn)[3]

DICOTYLEDONS	DICOTYLEDONS
Eritrichium (Boragineae)	Specularia (Campanulaceae)
Cuscuta (Convolvulaceae)	Campanula (Campanulaceae)
Scrophularia (Scrophularineae)	Hottonia (Primulaceae)
Linaria (Scrophularineae)	Anandria (Compositae)
Vandellia (Scrophularineae)	Heterocarpaea (Cruciferae)
Cryphiacanthus (Acanthaceae)	Viola (Violaceae)
Eranthemum (Acanthaceae)	Helianthemum (Cistineae)
Daedalacanthus (Acanthaceae)	Lechea (Cistineae)
Dipteracanthus (Acanthaceae)	Pavonia (Malvaceae)
Aechmanthera (Acanthaceae)	Gaudichaudia (Malpighiaceae)
Ruellia (Acanthaceae)	Aspicarpa (Malpighiaceae)
Lamium (Labiatae)	Camarea (Malpighiaceae)
Salvia (Labiatae)	Janusia (Malpighiaceae)
Oxybaphus (Nyctagineae) /	Polygala (Polygaleae)
Nyctaginia (Nyctagineae)	Impatiens (Balsamineae)
Stapelia (Asclepiadae)	Oxalis (Geraniaceae)

[3] I have omitted Trifolium and Arachis from the list, because Von Mohl says (*Bot. Zeitung*, 1863, p. 312) that the flower-stems merely draw the flowers beneath the ground, and that these do not appear to be properly cleistogamic. Correa de Mello (*Journal Linn. Soc. Bot.*, vol. xi, 1870, p. 254) observed plants of Arachis in Brazil, and could never find such flowers. Plantago has been omitted because as far as I can discover it produces hermaphrodite and female flower-heads, but not cleistogamic flowers. Krascheninikowia (vel Stellaria) has been omitted because it seems very doubtful from Maximowicz' description whether the lower flowers which have no petals or very small ones, and barren stamens or none, are cleistogamic; the upper hermaphrodite flowers are said never to produce fruit, and therefore probably act as males. Moreover in *Stellaria graminea*, as Babington remarks (*British Botany*, 1851, p. 51) 'shorter and longer petals accompany an imperfection of the stamens or germen'.

I have added to the list the following cases: Several Acanthaceae, for which see J. Scott in *Journal of Bot.* (London), new series, vol. i, 1872, p. 161. With respect to Salvia see Dr Ascherson in *Bot. Zeitung*, 1871, p. 555. For Oxybaphus and Nyctaginia see Asa Gray in *American Naturalist*, November, 1873, p. 692. From Dr Torrey's account of *Hottonia inflata* (*Bull. of Torrey Botan. Club*, vol. ii, June, 1871) it is manifest that this plant produces true cleistogamic flowers. For Pavonia see Bouché in *Sitzungsberichte d. Gesellsch. Natur. Freunde*, 20 October, 1874, p. 90. I have added Thelymitra, as from the account given by Mr Fitzgerald in his magnificent work on *Australian Orchids* it appears that the flowers of this plant in its native home never open, but they do not appear to be reduced in size. Nor is this the case with the flowers of certain species of Epidendron, Cattleya, etc. (see second edition of my *Fertilization of Orchids*, p. 147), which without expanding produce capsules. It is therefore doubtful whether these Orchideae ought to have been included in the list. From what Duval-Jouve says about Cryptostachys in *Bull. Soc. Bot. de France*, vol. x, 1863, p. 195, this plant appears to produce cleistogamic flowers. The other additions to the list are noticed in my text.

TABLE 38 – *continued*

DICOTYLEDONS	DICOTYLEDONS
Ononis (Leguminosae)	Drosera (Droseraceae)
Parochaetus (Leguminosae)	
Chapmannia (Leguminosae)	MONOCOTYLEDONS
Stylosanthus (Leguminosae)	Juncus (Junceae)
Lespedeza (Leguminosae)	Leersia (Gramineae)
Vicia (Leguminosae)	Hordeum (Gramineae)
Lathyrus (Leguminosae)	Cryptostachys (Gramineae)
Martinsia *vel*	Commelina (Commelineae)
Neurocarpum (Leguminosae)	Monochoria (Pontederaceae)
Amphicarpaea (Leguminosae)	Schomburgkia (Orchidae)
Glycine (Leguminosae)	Cattleya (Orchidae)
Galactia (Leguminosae)	Epidendron (Orchidae)
Voandzeia (Leguminosae)	Thelymitra (Orchidae)

more common in the family of the Leguminosae than in any other, and next in order in that of the Acanthaceae and Malpighiaceae. A large number, but not all the species, of certain genera, as of Oxalis and Viola, bear cleistogamic as well as ordinary flowers. A second point which deserves notice is that a considerable proportion of the genera produce more or less irregular flowers; this is the case with about 32 out of the 55 genera, but to this subject I shall recur.

I formerly made many observations on cleistogamic flowers, but only a few of them are worth giving, since the appearance of an admirable paper by Hugo von Mohl,[4] whose examination was in some respects much more complete than mine. His paper includes also an interesting history of our knowledge on the subject.

Viola canina. The calyx of the cleistogamic flowers differs in no respect from that of the perfect ones. The petals are reduced to five minute scales; the lower one, which represents the lower lip, is considerably larger than the others, but with no trace of the spur-like nectary; its margins are smooth, whilst those of the other four scale-like petals are papillose. D. Müller of Upsala says that in the specimens which he observed the petals were completely aborted.[5] The stamens are very small, and only the two lower ones are provided with anthers, which do not cohere together as in the perfect flowers. The anthers are minute,

---

[4] *Bot. Zeitung*, 1863, pp. 309–28.
[5] Ibid., 1857, p. 730. This paper contains the first full and satisfactory account of any cleistogamic flower.

with the two cells or loculi remarkably distinct; they contain very little pollen in comparison with those of the perfect / flowers. The connective expands into a membranous hood-like shield which projects above the anther-cells. These two lower stamens have no vestige of the curious appendages which secrete nectar in the perfect flowers. The three other stamens are destitute of anthers and have broader filaments, with their terminal membranous expansions flatter or not so hood-like as those of the two antheriferous stamens. The pollen grains have remarkably thin transparent coats; when exposed to the air they shrivel up quickly; when placed in water they swell, and are then 8–$10/7000$ of an inch in diameter, and therefore of smaller size than the ordinary pollen grains similarly treated, which have a diameter of 13–$14/7000$ of an inch. In the cleistogamic flowers, the pollen grains, as far as I could see, never naturally fall out of the anther-cells, but emit their tubes through a pore at the upper end. I was able to trace the tubes from the grains some way down the stigma. The pistil is very short, with the style hooked, so that its extremity, which is a little enlarged or funnel-shaped and represents the stigma, is directed downwards, being covered by the two membranous expansions of the antheriferous stamens. It is remarkable that there is an open passage from the enlarged funnel-shaped extremity to within the ovarium; this was evident, as slight pressure caused a bubble of air, which had been drawn in by some accident, to travel freely from one end to the other; a similar passage was observed by Michalet in *V. alba*. The pistil therefore differs considerably from that of the perfect flower; for in the latter it is much longer, and straight with the exception of the rectangularly bent stigma; nor is it perforated by an open passage.

The ordinary or perfect flowers have been said by some authors never to produce capsules; but this is an / error, though only a small proportion of them do so. This appears to depend in some cases on their anthers not containing even a trace of pollen, but more generally on bees not visiting the flowers. I twice covered with a net a group of flowers, and marked with threads twelve of them which had not as yet expanded. This precaution is necessary, for though as a general rule the perfect flowers appear considerably before the cleistogamic ones, yet occasionally some of the latter are produced early in the season, and their capsules might readily be mistaken for those produced by the perfect flowers. Not one of the twelve marked perfect flowers yielded a capsule, whilst others under the net which had been artificially fertilized produced five capsules; and these contained exactly the same average number of seeds as some capsules from flowers outside the net which had been fertilized by bees. I have repeatedly seen *Bombus hortorum, lapidarius*, and a third species, as well as hive-bees, sucking the flowers of this violet; I marked six which were thus visited, and four of them poduced fine capsules; the two others were gnawed off by some animal. I watched *Bombus hortorum* for some time, and whenever it came to a flower which did not stand in a convenient position to be sucked, it bit a hole through the spur-like nectary. Such ill-placed flowers would not yield any seed or leave descendants; and the plants bearing them would thus tend to be eliminated through Natural Selection.

The seeds produced by the cleistogamic and perfect flowers do not differ in appearance or number. On two occasions I fertilized several perfect flowers with pollen from other individuals, and afterwards marked some cleistogamic flowers on the same plants; and the result was that 14 capsules produced by the perfect / flowers contained on an average 9·85 seeds; and 17 capsules from the cleistogamic ones contained 9·64 seeds – an amount of difference of no significance. It is remarkable how much more quickly the capsules from the cleistogamic flowers are developed than those from the perfect ones; for instance several perfect flowers were cross-fertilized on 14 April, 1863, and a month afterwards (15 May) eight young cleistogamic flowers were marked with threads; and when the two sets of capsules thus produced were compared on 3 June, there was scarcely any difference between them in size.

Viola odorata (white-flowered, single, cultivated variety). The petals are represented by mere scales as in the last species; but differently from in the last, all five stamens are provided with diminutive anthers. Small bundles of pollen tubes were traced from the five anthers into the somewhat distant stigma. The capsules produced by these flowers bury themselves in the soil, if it be loose enough, and there mature themselves.[6] Lecoq says that it is only these latter capsules which possess elastic valves; but I think this must be a misprint, as such valves would obviously be of no use to the buried capsules, but would serve to scatter the seeds of the sub-aerial ones, as in the other species of Viola. It is remarkable that this plant, according to Delpino,[7] does not produce

---

[6] Vaucher says (*Hist. Phys. des Plantes d'Europe*, vol. iii, 1844, p. 309) that *V. hirta* and *collina* likewise bury their capsules. See also Lecoq, *Géograph. Bot.*, vol. v, 1856, p. 180.

[7] *Sull' Opera, la Distribuzione dei Sessi nelle Piante*, etc., 1867, p. 30.

cleistogamic flowers in one part of Liguria, whilst the perfect flowers are there abundantly fertile; on the other hand, cleistogamic flowers are produced by it near Turin. Another fact is worth giving as an instance of correlated / development; I found on a purple variety, after it had produced its perfect double flowers, and whilst the white single variety was bearing its cleistogamic flowers, many bud-like bodies which from their position on the plant were certainly of a cleistogamic nature. They consisted, as could be seen on bisecting them, of a dense mass of minute scales closely folded over one another, exactly like a cabbage-head in miniature. I could not detect any stamens, and in the place of the ovarium there was a little central column. The doubleness of the perfect flowers had thus spread to the cleistogamic ones, which therefore were rendered quite sterile.

Viola hirta. The five stamens of the cleistogamic flowers are provided, as in the last case, with small anthers, from all of which pollen tubes proceed to the stigma. The petals are not quite so much reduced as in *V. canina,* and the short pistil instead of being hooked is merely bent into a rectangle. Of several perfect flowers which I saw visited by hive- and humble-bees, six were marked, but they produced only two capsules, some of the others having been accidentally injured. M. Monnier was therefore mistaken in this case as in that of *V. odorata,* in supposing that the perfect flowers always withered away and aborted. He states that the peduncles of the cleistogamic flowers curve downwards and bury the ovaries beneath the soil.[8] I may here add that Fritz Müller, as I hear from his brother, has found in the highlands of Southern Brazil a white-flowered species of violet which bears sub-terranean cleistogamic flowers. /

Viola nana. Mr Scott sent me seeds of this Indian species from the Sikkim Terai, from which I raised many plants, and from these other seedlings during several successive generations. They produced an abundance of cleistogamic flowers during the whole of each summer, but never a perfect one. When Mr Scott wrote to me his plants in Calcutta were behaving similarly, though his collector saw the species in flower in its native site. This case is valuable as showing that we ought not to infer, as has sometimes been done, that a species does not

[8] These statements are taken from Professor Oliver's excellent article in the *Nat. Hist. Review,* July, 1862, p. 238. With respect to the supposed sterility of the perfect flowers in this genus see also Timbal-Lagrave in *Bot. Zeitung,* 1854, p. 772.

bear perfect flowers when growing naturally, because it produces only cleistogamic flowers under culture. The calyx of these flowers is sometimes formed of only three sepals; two being actually suppressed and not merely coherent with the others; this occurred with five out of thirty flowers which were examined for this purpose. The petals are represented by extremely minute scales. Of the stamens, two bear anthers which are in the same state as in the previous species, but, as far as I could judge, each of the two cells contained only from 20 to 25 delicate transparent pollen grains. These emitted their tubes in the usual manner. The three other stamens bore very minute rudimentary anthers, one of which was generally larger than the other two, but none of them contained any pollen. In one instance, however, a single cell of the larger rudimentary anther included a little pollen. The style consists of a short flattened tube, somewhat expanded at its upper end, and this forms an open channel leading into the ovarium, as described under *V. canina*. It is slightly bent towards the two fertile anthers.

Viola Roxburghiana. This species bore in my hot-house during two years a multitude of cleistogamic flowers, which resembled in all respects those of the / last species; but no perfect ones were produced. Mr Scott informs me that in India it bears perfect flowers only during the cold season, and that these are quite fertile. During the hot, and more especially during the rainy season, it bears an abundance of cleistogamic flowers.

Many other species, besides the five now described, produce cleistogamic flowers; this is the case, according to D. Müller, Michalet, Von Mohl, and Hermann Müller, with *V. elatior, lancifolia, sylvatica, palustris, mirabilis, bicolor, ionodium,* and *biflora.* But *V. tricolor* does not produce them.

Michalet asserts that *V. palustris* produces near Paris only perfect flowers, which are quite fertile; but that when the plant grows on mountains cleistogamic flowers are produced; and so it is with *V. biflora.* The same author states that he has seen in the case of *V. alba* flowers intermediate in structure between the perfect and cleistogamic ones. According to M. Boisduval, an Italian species, *V. Ruppii,* never bears in France 'des fleurs bien apparentes, ce qui ne l'empêche pas de fructifier'.

It is interesting to observe the gradation in the abortion of the parts in the cleistogamic flowers of the several foregoing species. It appears from the statements by D. Müller and Von Mohl that in *V. mirabilis* the calyx does not remain quite closed; all five stamens are provided with anthers, and some pollen grains probably fall out of the cells on the stigma, instead of protruding their tubes whilst still enclosed, as in the other species. In *V. hirta* all five stamens are likewise antheriferous; the petals are not so much reduced and the pistil not so much modified as in the following species. In *V. nana* and *elatior* only two of the stamens properly bear anthers, but / sometimes one or even two of the others are thus provided. Lastly, in *V. canina* never more than two of the stamens, as far as I have seen, bear anthers; the petals are much more reduced than in *V. hirta,* and according to D. Müller are sometimes quite absent.

Oxalis acetosella. The existence of cleistogamic flowers on this plant was discovered by Michalet.[9] They have been fully described by Von Mohl, and I can add hardly anything to his description. In my specimens the anthers of the five longer stamens were nearly on a level with the stigmas; whilst the smaller and less plainly bilobed anthers of the five shorter stamens stood considerably below the stigmas, so that their tubes had to travel some way upwards. According to Michalet these latter anthers are sometimes quite aborted. In one case the tubes, which ended in excessively fine points, were seen by me stretching upwards from the lower anthers towards the stigmas, which they had not as yet reached. My plants grew in pots, and long after the perfect flowers had withered they produced not only cleistogamic but a few minute open flowers, which were in an intermediate condition between the two kinds. In one of these the pollen tubes from the lower anthers had reached the stigmas, though the flower was open. The footstalks of the cleistogamic flowers are much shorter than those of the perfect flowers, and are so much bowed downwards that they tend, according to Von Mohl, to bury themselves in the moss and dead leaves on the ground. Michalet also says that they are often hypogean. In order to ascertain the number of seeds produced by these flowers, I marked eight of them; two failed, one cast its seed abroad, and the / remaining five contained on an average 10·0 seeds per capsule. This is rather above the average 9·2, which eleven capsules from perfect flowers fertilized with their own pollen yielded, and considerably above the average 7·9, from the capsules of perfect flowers fertilized with pollen from another plant; but this latter result must, I think, have been accidental.

[9] *Bull. Soc. Bot. de France,* vol. vii, 1860, p. 465.

Hildebrand, whilst searching various Herbaria, observed that many other species of Oxalis besides *O. acetosella* produce cleistogamic flowers;[10] and I hear from him that this is the case with the heterostyled trimorphic *O. incarnata* from the Cape of Good Hope.

Oxalis (Biophytum) sensitiva. This plant is ranked by many botanists as a distinct genus, but as a subgenus by Bentham and Hooker. Many of the early flowers on a mid-styled plant in my hot-house did not open properly, and were in an intermediate condition between cleistogamic and perfect. Their petals varied from a mere rudiment to about half their proper size; nevertheless they produced capsules. I attributed their state to unfavourable conditions, for later in the season fully expanded flowers of the proper size appeared. But Mr Thwaites afterwards sent me from Ceylon a number of long-styled, mid-styled, and short-styled flower-stalks preserved in spirits; and on the same stalks with the perfect flowers, some of which were fully expanded and others still in bud, there were small bud-like bodies containing mature pollen, but with their calyces closed. These cleistogamic flowers do not differ much in structure from the perfect ones of the corresponding form, with the exception that their petals are reduced to extremely minute, barely visible scales, which adhere firmly to the rounded / bases of the shorter stamens. Their stigmas are much less papillose, and smaller in about the ratio of 13 to 20 divisions of the micrometer, as measured transversely from apex to apex, than the stigmas of the perfect flowers. The styles are furrowed longitudinally, and are clothed with simple as well as glandular hairs, but only in the cleistogamic flowers produced by the long-styled and mid-styled forms. The anthers of the longer stamens are a little smaller than the corresponding ones of the perfect flowers, in about the ratio of 11 to 14. They dehisce properly, but do not appear to contain much pollen. Many pollen grains were attached by short tubes to the stigmas; but many others, still adhering to the anthers, had emitted their tubes to a considerable length, without having come in contact with the stigmas. Living plants ought to be examined, as the stigmas, at least of the long-styled form, project beyond the calyx, and if visited by insects (which, however, is very improbable) might be fertilized with pollen from a perfect flower. The most singular fact about the present species is that long-styled cleistogamic flowers are produced by the long-styled

[10] *Monatsbericht der Akad. der Wiss. zu Berlin*, 1866, p. 369.

plants, and mid-styled as well as short-styled cleistogamic flowers by
the other two forms, so that there are three kinds of cleistogamic and
three kinds of perfect flowers produced by this one species! Most of
the heterostyled species of Oxalis are more or less sterile, many
absolutely so, if illegitimately fertilized with their own-form pollen. It
is therefore probable that the pollen of the cleistogamic flowers has
been modified in power, so as to act on their own stigmas, for they
yield an abundance of seeds. We may perhaps account for the cleistogamic flowers consisting of the three forms, through the principle of
correlated growth, by which the cleistogamic / flowers of the double
violet have been rendered double.

Vandellia nummularifolia. Dr Kuhn has collected all the notices with
respect to cleistogamic flowers in this genus, and has described from
dried specimens those produced by an Abyssinian species. Mr Scott
sent me from Calcutta seeds of the above common Indian weed, from
which many plants were successively raised during several years. The
cleistogamic flowers are very small, being when fully mature under $\frac{1}{20}$
of an inch (1·27 mm) in length. The calyx does not open, and within it
the delicate transparent corolla remains closely folded over the
ovarium. There are only two anthers instead of the normal number of
four, and their filaments adhere to the corolla. The cells of the anthers
diverge much at their lower ends and are only $\frac{5}{700}$ of an inch (0·181
mm) in their longer diameter. They contain but few pollen grains, and
these emit their tubes whilst still within the anther. The pistil is very
short, and is surmounted by a bilobed stigma. As the ovary grows the
two anthers together with the shrivelled corolla, all attached by the
dried pollen tubes to the stigma, are torn off and carried upwards in
the shape of a little cap. The perfect flowers generally appear before
the cleistogamic, but sometimes simultaneously with them. During one
season a large number of plants produced no perfect flowers. It has
been asserted that the latter never yield capsules; but this is a mistake,
as they do so even when insects are excluded. Fifteen capsules from
cleistogamic flowers on plants growing under favourable conditions
contained on an average 64·2 seeds, with a maximum of 87; whilst 20
capsules from plants growing much / crowded yielded an average of
only 48. Sixteen capsules from perfect flowers artificially crossed with
pollen from another plant contained on an average 93 seeds, with a

[11] *Bot. Zeitung*, 1867, p. 65.

maximum of 137. Thirteen capsules from self-fertilized perfect flowers gave an average of 62 seeds, with a maximum of 135. Therefore the capsules from the cleistogamic flowers contained fewer seeds than those from perfect flowers when cross-fertilized, and slightly more than those from perfect flowers self-fertilized.

Dr Kuhn believes that the Abyssinian *V. sessiflora* does not differ specifically from the foregoing species. But its cleistogamic flowers apparently include four anthers instead of two as above described. The plants, moreover, of *V. sessiflora* produce subterranean runners which yield capsules; and I never saw a trace of such runners in *V. nummularifolia*, although many plants were cultivated.

Linaria spuria. Michalet says[12] that short, thin, twisted branches are developed from the buds in the axils of the lower leaves, and that these bury themselves in the ground. They there produce flowers not offering any peculiarity in structure, excepting that their corollas, though properly coloured, are deformed. These flowers may perhaps be ranked as cleistogamic, as they are developed, and not merely drawn, beneath the ground.

Ononis columnae. Plants were raised from seeds sent me from Northern Italy. The sepals of the cleistogamic flowers are elongated and closely pressed together; the petals are much reduced in size, colourless, and folded over the interior organs. The filaments of the ten stamens are united into a tube, and / this is not the case, according to Von Mohl, with the cleistogamic flowers of other Leguminosae. Five of the stamens are destitute of anthers, and alternate with the five thus provided. The two cells of the anthers are minute, rounded and separated from one another by connective tissue; they contain but few pollen grains, and these have extremely delicate coats. The pistil is hook-shaped, with a plainly enlarged stigma, which is curled down, towards the anthers; it therefore differs much from that of the perfect flower. During the year 1867 no perfect flowers were produced, but in the following year there were both perfect and cleistogamic ones.

Ononis minutissima. My plants produced both perfect and cleistogamic flowers; but I did not examine the latter. Some of the former were crossed with pollen from a distinct plant, and six capsules thus

---

[12] *Bull. Soc. Bot. de France*, vol. vii, 1860, p. 468.

obtained yielded on an average 3·66 seeds, with a maximum of 5 in one. Twelve perfect flowers were marked and allowed to fertilize themselves spontaneously under a net, and they yielded eight capsules, containing on an average 2·38 seeds, with a maximum of 3 in one. Fifty-three capsules produced by the cleistogamic flowers contained on an average 4·1 seeds, so that these were the most productive of all; and the seeds themselves looked finer even than those from the crossed perfect flowers. According to Mr Bentham *O. parviflora* likewise bears cleistogamic flowers; and he informs me that these flowers are produced by all three species early in the spring; whilst the perfect ones appear afterwards, and therefore in a reversed order compared with those of Viola and Oxalis. Some of the species, for instance *Ononis columnae*, bear a fresh crop of cleistogamic flowers in the autumn.

Lathyrus nissolia apparently offers a case of the first / stage in the production of cleistogamic flowers, for on plants growing in a state of nature, many of the flowers never expand and yet produce fine pods. Some of the buds are so large that they seem on the point of expansion; others are much smaller, but none so small as the true cleistogamic flowers of the foregoing species. As I marked these buds with thread and examined them daily, there could be no mistake about their producing fruit without having expanded.

Several other Leguminous genera produce cleistogamic flowers, as may be seen in the previous list; but much does not appear to be known about them. Von Mohl says that their petals are commonly rudimentary, that only a few of their anthers are developed, their filaments are not united into a tube and their pistils are hook-shaped. In three of the genera, namely Vicia, Amphicarpaea, and Voandzeia, the cleistogamic flowers are produced on subterranean stems. The perfect flowers of Voandzeia, which is a cultivated plant, are said never to produce fruit;[13] but we should remember how often fertility is affected by cultivation.

Impatiens fulva. Mr A. W. Bennett has published an excellent description, with figures, of this plant.[14] He shows that the cleistogamic and perfect flowers differ in structure at a very early period of growth, so that the existence of the former cannot be due merely to the arrested

[13] Correa de Mello (*Journal Linn. Soc. Bot.*, vol. xi, 1870, p. 254) particularly attended to the flowering and fruiting of this African plant, which is sometimes cultivated in Brazil.

[14] *Journal Linn. Soc. Bot.*, vol. xiii, 1872, p. 147.

development of the latter – a conclusion which indeed follows from most of the previous descriptions. Mr Bennett found on the banks of the Wey that the plants which bore cleistogamic flowers alone were to those bearing perfect flowers as 20 to 1, but / we should remember that this is a naturalized species. The perfect flowers are usually barren in England; but Professor Asa Gray writes to me that after midsummer in the United States some or many of them produce capsules.

Impatiens noli-me-tangere. I can add nothing of important to Von Mohl's description, excepting that one of the rudimentary petals shows a vestige of a nectary, as Mr Bennett likewise found to be the case with *I. fulva*. As in this latter species all five stamens produce some pollen, though small in amount; a single anther contains, according to Von Mohl, not more than 50 grains, and these emit their tubes while still enclosed within it. The pollen grains of the perfect flowers are tied together by threads, but not, so far as I could see, those of the cleisto-gamic flowers; and a provision of this kind would here have been useless, as the grains can never be transported by insects. The flowers of *I. balsamina* are visited by humble-bees,[15] and I am almost sure that this is the case with the perfect flowers of *I. noli-me-tangere*. From the perfect flowers of this latter species covered with a net eleven sponta-neously self-fertilized capsules were produced, and these yielded on an average 3·45 seeds. Some perfect flowers with their anthers still con-taining an abundance of pollen were fertilized with pollen from a distinct plant; and the three capsules thus produced contained, to my surprise, only 2, 2, and 1 seed. As *I. balsamina* is proterandrous, so probably is the present species; and if so, cross-fertilization was affected by me at too early a period, and this may account for the capsules yielding so few seeds.

Drosera rotundifolia. The first flower-stems which / were thrown up by some plants in my greenhouse bore only cleistogamic flowers. The petals of small size remained permanently closed over the reproduc-tive organs, but their white tips could just be seen between the almost completely closed sepals. The pollen, which was scanty in amount, but not so scanty as in Viola or Oxalis, remained enclosed within the anthers, whence the tubes proceeded and penetrated the stigma. As the ovarium swelled the little withered corolla was carried upwards in

[15] H. Müller, *Die Befruchtung*, etc., p. 170.

the form of a cap. These cleistogamic flowers produced an abundance of seed. Later in the season perfect flowers appeared. With plants in a state of nature the flowers open only in the early morning, as I have been informed by Mr Wallis, who particularly attended to the time of their flowering. In the case of *D. Anglica*, the still folded petals on some plants in my greenhouse opened just sufficiently to leave a minute aperture; the anthers dehisced properly, but the pollen grains adhered in a mass to them, and thence emitted their tubes, which penetrated the stigmas. These flowers, therefore, were in an intermediate condition, and could not be called either perfect or cleistogamic.

A few miscellaneous observations may be added with respect to some other species, as throwing light on our subject. Mr Scott states[16] that *Eranthemum ambiguum* bears three kinds of flowers – large, conspicuous, open ones, which are quite sterile – others of intermediate size, which are open and moderately fertile – and lastly small closed or cleistogamic ones which are perfectly fertile. *Ruellia tuberosa*, likewise one of the Acanthaceae, produces both open and cleistogamic / flowers; the latter yield from 18 to 24, whilst the former only from 8 to 10 seeds; these two kinds of flowers are produced simultaneously, whereas in several other members of the family the cleistogamic ones appear only during the hot season. Acccording to Torrey and Gray, the North American species of Helianthemum, when growing in poor soil, produce only cleistogamic flowers. The cleistogamic flowers of *Specularia perfoliata* are highly remarkable, as they are closed by a tympanum formed by the rudimentary corolla, and without any trace of an opening. The stamens vary from 3 to 5 in number, as do the sepals.[17] The collecting hairs on the pistil, which play so important a part in the fertilization of the perfect flowers are here quite absent. Sir J. Hooker and Dr Thomson state[18] that some of the Indian species of Campanula produce two kinds of flowers; the smaller ones being borne on longer peduncles with differently formed sepals, and producing a more globose ovary. The flowers are closed by a tympanum like that in Specularia. Some of the plants produce both kinds of flowers, others only one kind; both yield an abundance of

[16] *Journal of Botany*, London, new series, vol. i, 1872, pp. 161–4.

[17] Von Mohl, *Bot. Zeitung*, 1863, pp. 314 and 323. Dr Bromfield (*Phytologist*, vol. iii, p. 530) also remarks that the calyx of the cleistogamic flowers is usually only 3-cleft, while that of the perfect flower is mostly 5-cleft.

[18] *Journal Linn. Soc.*, vol. ii, 1857, p. 7. See also Professor Oliver in *Nat. Hist. Review*, 1862, p. 240.

seeds. Professor Oliver adds that he has seen flowers on *Campanula colorata* in an intermediate condition between cleistogamic and perfect ones.

The solitary almost sessile cleistogamic flowers produced by *Monochoria vaginalis* are differently protected from those in any of the previous cases, namely within 'a short sack formed of the membranous spathe, without any opening or fissure'. There is only a single / fertile stamen; the style is almost obsolete, with the three stigmatic surfaces directed to one side. Both the perfect and cleistogamic flowers produce seeds.[19]

The cleistogamic flowers on some of the Malpighaceae seem to be more profoundly modified than those in any of the foregoing genera. According to A. de Jussieu[20] they are differently situated from the perfect flowers; they contain only a single stamen, instead of 5 or 6; and it is a strange fact that this particular stamen is not developed in the perfect flowers of the same species. The style is absent or rudimentary; and there are only two ovaries instead of three. Thus these degraded flowers, as Jussieu remarks, 'laugh at our classifications, for the greater number of the characters proper to the species, to the genus, to the family, to the class disappear'. The calyces of the perfect flowers are studded with glands, and their absence on the cleistogamic flowers may probably be explained by an observation of Fritz Müller, who informs me that in the one species, *Bunchosia Gaudichaudiana*, the fertilization of which he has often witnessed, the perfect flowers are regularly visited by bees belonging to the genera Tetrapedia and Epicharis. These bees sit down on the flowers, gnawing the glands on the outside of the calyx, and in doing so the under sides of their bodies are dusted with pollen, by which afterwards other flowers are fertilized. Such visits to the cleistogamic flowers would be useless.

As the Asclepiadous genus Stapelia is said to produce cleistogamic flowers, the following case may be worth giving. I have never heard of the perfect flowers of *Hoya carnosa* setting seeds in this country, but some / capsules were produced in Mr Farrer's hot-house; and the gardener detected that they were the product of minute bud-like bodies, three or four of which could sometimes be found on the same umbel with the perfect flowers. They were quite closed and hardly

[19] Dr Kirk, *Journ. Linn. Soc.*, vol. viii, 1864, p. 147.
[20] *Archives du Muséum*, vol. iii, 1843, pp. 35–8, 82–6, 589, 598.

thicker than their peduncles. The sepals presented nothing particular, but internally and alternating with them, there were five small flattened heart-shaped papillae, like rudiments of petals; but the homological nature of which appeared doubtful to Mr Bentham and Dr Hooker. No trace of anthers or of stamens could be detected; and I knew from having examined many cleistogamic flowers what to look for. There were two ovaries, full of ovules, quite open at their upper ends, with their edges festooned, but with no trace of a proper stigma. In all these flowers one of the two ovaries withered and blackened long before the other. The one perfect capsule, 3½ inches in length which was sent me, had likewise been developed from a single carpel. This capsule contained an abundance of plumose seeds, many of which appeared quite sound, but they did not germinate when sown at Kew. Therefore the little bud-like flower which produced this capsule probably was as destitute of pollen as were those which I examined.

Juncus bufonius and Hordeum. All the species hitherto mentioned which produce cleistogamic flowers are entomophilous; but Juncus and seven genera of Gramineae are anemophilous. *Juncus bufonius* is remarkable[21] by bearing in parts of Russia only cleistogamic flowers, which contain three instead of the six anthers found in the perfect flowers. In the / genus *Hordeum* it has been shown by Delpino[22] that the majority of the flowers are cleistogamic, some of the others expanding and apparently allowing of cross-fertilization. I hear from Fritz Müller that there is a grass in Southern Brazil, in which the sheath of the uppermost leaf, half a metre in length, envelopes the whole panicle; and this sheath never opens until the self-fertilized seeds are ripe. On the roadside some plants had been cut down, whilst the cleistogamic panicles were developing, and these plants afterwards produced free or unenclosed panicles of small size, bearing perfect flowers.

Leersia oryzoides. It has long been known that this plant produces cleistogamic flowers, but these were first described with care by M. Duval-Jouve.[23] I procured plants from a stream near Reigate, and

[21] See Dr Ascherson's interesting paper in *Bot. Zeitung*, 1871, p. 551. Also 1872, p. 697.
[22] *Bollettini del Comizio agrario Parmense*. Marzo e Aprile, 1871. An abstract of this valuable paper is given in *Bot. Zeitung*, 1871, p. 537. See also Hildebrand on Hordeum, in *Monatsbericht d. K. Akad. Berlin*, October, 1872, p. 760.
[23] *Bull. Bot. Soc. de France*, vol. x, 1863, p. 194.

cultivated them for several years in my greenhouse. The cleistogamic flowers are very small, and usually mature their seeds within the sheaths of the leaves. These flowers are said by Duval-Jouve to be filled by slightly viscid fluid; but this was not the case with several that I opened; but there was a thin film of fluid between the coats of the glumes, and when these were pressed the fluid moved about, giving a singularly deceptive appearance of the whole inside of the flower being thus filled. The stigma is very small and the filaments extremely short; the anthers are less than 1/50 of an inch in length, or about one-third of the length of those in the perfect flowers. One of the three anthers dehisces before the two others. Can this have any relation with the fact that in some other / species of Leersia only two stamens are fully developed?[24] The anthers shed their pollen on the stigma; at least in one instance this was clearly the case, and by tearing open the anthers under water the grains were easily detached. Towards the apex of the anther the grains are arranged in a single row and lower down in two or three rows, so that they could be counted; and there were about 35 in each cell, or 70 in the whole anther; and this is an astonishingly small number for an anemophilous plant. The grains have very delicate coats, are spherical and about 5/7000 of an inch (0·0181 mm), whilst those of the perfect flowers are about 7/7000 of an inch (0·0254 mm) in diameter.

M. Duval-Jouve states that the panicles very rarely protrude from their sheaths, but that when this does happen the flowers expand and exhibit well-developed ovaries and stigmas, together with full-sized anthers containing apparently sound pollen; nevertheless such flowers are invariably quite sterile. Schreiber had previously observed that if a panicle is only half protruded, this half is sterile, whilst the still included half is fertile. Some plants which grew in a large tub of water in my greenhouse behaved on one occasion in a very different manner. They protruded two very large much-branched panicles; but the florets never opened, though these included fully developed stigmas and stamens supported on long filaments with large anthers that dehisced properly. If these florets had opened for a short time unperceived by me and had then closed again, the empty anthers would have been left dangling outside. Nevertheless they yielded on 17 August an abundance of fine ripe seeds. Here then we have a near approach to the / single

[24] Asa Gray, *Manual of Bot. of United States*, 1856, p. 540.

case as yet known[25] of this grass producing in a state of nature (in Germany) perfect flowers which yielded a copious supply of fruit. Seeds from the cleistogamic flowers were sent by me to Mr Scott in Calcutta, who there cultivated the plants in various ways, but they never produced perfect flowers.

In Europe *Leersia oryzoides* is the sole representative of its genus, and Duval-Jouve, after examining several exotic species, found that it apparently is the sole one which bears cleistogamic flowers. It ranges from Persia to North America, and specimens from Pennsylvania resembled the European ones in their concealed manner of fructification. There can therefore be little doubt that this plant generally propagates itself throughout an immense area by cleistogamic seeds, and that it can hardly ever be invigorated by cross-fertilization. It resembles in this respect those plants which are now widely spread though they increase solely by a sexual generation.[26]

Concluding remarks on cleistogamic flowers

That these flowers owe their structure primarily to the arrested development of perfect ones, we may infer from such cases as that of the lower rudimentary petal in Viola being larger than the others, like the lower lip of the perfect flower – from a vestige of a spur in the cleistogamic flowers of Impatiens – from the ten stamens of Ononis being united into a tube – and other such structures. The same inference may be drawn from the occurrence, in some instances, on the same plant of a series of gradations between the cleistogamic and perfect flowers. But that the former owe their origin wholly to arrested development is / by no means the case; for various parts have been specially modified, so as to aid in the self-fertilization of the flowers, and as a protection to the pollen; for instance, the hook-shaped pistil in Viola and in some other genera, by which the stigma is brought close to the fertile anthers – the rudimentary corolla of Specularia modified into a perfectly closed tympanum, and the sheath of Monochoria modified into a closed sack – the excessively thin coats of the pollen grains – the anthers not being all equally aborted, and other such cases. Moreover Mr Bennett has shown that the buds of the cleistogamic and perfect flowers of Impatiens differ at a very early period of growth.

[25] Dr Ascherson, *Bot. Zeitung*, 1864, p. 350.
[26] I have collected several such cases in my *Variation under Domestication*, ch. xviii, 2nd edit., vol. ii, p. 153.

The degree to which many of the most important organs in these degraded flowers have been reduced or even wholly obliterated, is one of their most remarkable peculiarities, reminding us of many parasitic animals. In some cases only a single anther is left, and this contains but few pollen grains of diminished size; in other cases the stigma has disappeared leaving a simple open passage into the ovarium. It is also interesting to note the complete loss of trifling points in the structure or functions of certain parts, which though of service to the perfect flowers, are of none to the cleistogamic; for instance the collecting hairs on the pistil of Specularia, the glands on the calyx of the Malpighiaceae, the nectar-secreting appendages to the lower stamens of Viola, the secretion of nectar by other parts, the emission of a sweet odour, and apparently the elasticity of the valves in the buried capsules of *Viola odorata*. We here see, as throughout nature, that as soon as any part or character becomes superfluous it tends sooner or later to disappear.

Another peculiarity in these flowers is that the / pollen grains generally emit their tubes whilst still enclosed within the anthers; but this is not so remarkable a fact as was formerly thought, when the case of Asclepias was alone known.[27] It is, however, a wonderful sight to behold the tubes directing themselves in a straight line to the stigma, when this is at some little distance from the anthers. As soon as they reach the stigma or the open passage leading into the ovarium, no doubt they penetrate it, guided by the same means, whatever these may be, as in the case of ordinary flowers. I thought that they might be guided by the avoidance of light: some pollen grains of a willow were therefore immersed in an extremely weak solution of honey, and the vessel was placed so that the light entered only in one direction, laterally or from below or from above, but the long tubes were in each case protruded in every possible direction.

As cleistogamic flowers are completely closed they are necessarily self-fertilized, not to mention the absence of any attraction to insects: and they thus differ widely from the great majority of ordinary flowers.

---

[27] The case of Asclepias was described by R. Brown. Baillon asserts (*Adansonia*, vol. ii, 1862, p. 58) that with many plants the tubes are emitted from pollen grains which have not come into contact with the stigma; and that they may be seen advancing horizontally through the air towards the stigma. I have observed the emission of the tubes from the pollen masses whilst still within the anthers, in three widely distinct Orchidean genera, namely Aceras, Malaxis, and Neottia: see *The Various Contrivances by which Orchids are Fertilised*, 2nd edit., p. 258.

Delpino believes[28] that cleistogamic flowers have been developed in order to ensure the production of seeds under climatic or other conditions which tend / to prevent the fertilization of the perfect flowers. I do no doubt that this holds good to a certain limited extent, but the production of a large supply of seeds with little consumption of nutrient matter or expenditure of vital force is probably a far more efficient motive power. The whole flower is much reduced in size: but what is much more important, an extremely small quantity of pollen has to be formed, as none is lost through the action of insects or the weather; and pollen contains much nitrogen and phosphorus. Von Mohl estimated that a single cleistogamic anther-cell of *Oxalis acetosella* contained from one to two dozen pollen grains; we will say 20, and if so the whole flower can have produced at most 400 grains; with Impatiens the whole number may be estimated in the same manner at 250; with Leersia at 210; and with *Viola nana* at only 100. These figures are wonderfully low compared with the 243,600 pollen grains produced by a flower of Leontodon, the 4,863 by an Hibiscus, or the 3,654,000 by a Paeony.[29] We thus see that heterostyled flowers produce seeds with a wonderfully small expenditure of pollen; and they produce as a general rule quite as many seeds as the perfect flowers.

That the production of a large number of seeds is necessary or beneficial to many plants needs no evidence. So of course is their preservation before they are ready for germination; and it is one of the many remarkable peculiarities of the plants which bear heterostyled flowers, that an incomparably larger proportion of them than of ordinary plants bury their young ovaries in the ground; an action which it may be presumed serves to protect them from being / devoured by birds or other enemies. But this advantage is accompanied by the loss of the power of wide dissemination. No less than eight of the genera in the list at the beginning of this chapter include species which act in this manner, namely, several kinds of Viola, Oxalis, Vandellia, Linaria, Commelina, and at least three genera of Leguminosae. The seeds also of Leersia, though not buried, are concealed in the most perfect manner within the sheaths of the leaves. Cleistogamic flowers possess great facilities for burying their young ovaries or capsules, owing to their small size, pointed shape, closed condition and

[28] *Sull' Opera la Distribuzione dei Sessa nelle Piante,* 1867, p. 30.
[29] The authorities for these statements are given in my *Effects of Cross and Self-Fertilisation,* p. 376.

the absence of a corolla; and we can thus understand how it is that so many of them have acquired this curious habit.

It has already been shown that in about 33 out of the 67 genera in the list just referred to, the perfect flowers are irregular; and this implies that they have been specially adapted for fertilization by insects. Moreover three of the genera with regular flowers are adapted by other means for the same end. Flowers thus constructed are liable during certain seasons to be imperfectly fertilized, namely, when the proper insects are scarce; and it is difficult to avoid the belief that the production of cleistogamic flowers, which ensures under all circumstances a fully supply of seed, has been in part determined by the perfect flowers being liable to fail in their fertilization. But if this determining cause be a real one, it must be of subordinate importance, as eight of the genera in the list are fertilized by the wind; and there seems no reason why their perfect flowers should fail to be fertilized more frequently than those in any other anemophilous genus. In contrast with what we here see with respect to the large proportion of the perfect / flowers being irregular, one genus alone out of the 38 heterostyled genera described in the previous chapters bears such flowers; yet all these genera are absolutely dependent on insects for their legitimate fertilization. I know not how to account for this difference in the proportion of the plants bearing regular and irregular flowers in the two classes, unless it be that the heterostyled flowers are already so well adapted for cross-fertilization, through the position of their stamens and pistils and the difference in power of their two or three kinds of pollen, that any additional adaptation, namely, through the flowers being made irregular, has been rendered superfluous.

Although cleistogamic flowers never fail to yield a large number of seeds, yet the plants bearing them usually produce perfect flowers, either simultaneously or more commonly at a different period; and these are adapted for or admit of cross-fertilization. From the cases given of the two Indian species of Viola, which produced in this country during several years only cleistogamic flowers, and of the numerous plants of Vandellia and of some plants of Ononis which behaved during one whole season in the same manner, it appears rash to infer from such cases as that of *Salvia cleistogama* not having produced perfect flowers during five years in Germany,[30] and of an

[30] Dr Ascherson, *Bot. Zeit.*, 1871, p. 555.

Aspicarpa not having done so during several years in Paris, that these plants would not bear perfect flowers in their native homes. Von Mohl and several other botanists have repeatedly insisted that as a general rule the perfect flowers produced by cleistogamic plants are sterile; but it has been shown under the head of the several species that this is not the case. The perfect / flowers of Viola are indeed sterile unless they are visited by bees; but when thus visited they yield the full number of seeds. As far as I have been able to discover there is only one absolute exception to the rule that the perfect flowers are fertile, namely, that of Voandzeia; and in this case we should remember that cultivation often affects injuriously the reproductive organs. Although the perfect flowers of Leersia sometimes yield seeds, yet this occurs so rarely, as far as hitherto observed, that it practically forms a second exception to the rule.

As cleistogamic flowers are invariably fertilized, and as they are produced in large numbers, they yield altogether a much larger supply of seeds than do the perfect flowers on the same plant. But the latter flowers will occasionally be cross-fertilized, and their offspring will thus be invigorated, as we may infer from a widespread analogy. But of such invigoration I have only a small amount of direct evidence: two crossed seedlings of *Ononis minutissima* were put into competition with two seedlings raised from cleistogamic flowers; they were at first all of equal height; the crossed were then slightly beaten; but on the following year they showed the usual superiority of their class, and were to the self-fertilized plants of cleistogamic origin as 100 to 88 in mean height. With Vandellia twenty crossed plants exceeded in height twenty plants raised from cleistogamic seeds only by a little, namely, in the ratio of 100 to 94.

It is a natural enquiry how so many plants belonging to various very distinct families first came to have the development of their flowers arrested, so as ultimately to become cleistogamic. That a passage from the one state to the other is far from difficult is shown by the many recorded cases of gradations between the / two states on the same plant, in Viola, Oxalis, Biophytum, Campanula, etc. In the several species of Viola the various parts of the flowers have also been modified in very different degrees. Those plants which in their own country produce flowers of full or nearly full size, but never expand (as with Thelymitra), and yet set fruit, might easily be rendered cleistogamic. *Lathyrus nissolia* seems to be in an incipient transitional state, as does *Drosera Anglica*, the flowers of which are not perfectly closed. There is good evidence that flowers sometimes fail to expand and are somewhat reduced in size, owing to exposure to unfavourable conditions, but still retain their fertility unimpaired. Linnaeus observed in 1763 that the flowers on several plants brought from Spain and grown at Upsala did not show any corolla and yet produced seeds. Asa Gray has seen flowers on exotic plants in the Northern United States which never expanded and yet fruited. With certain English plants, which bear flowers during nearly the whole year, Mr Bennett found that those produced during the winter season were fertilized in the bud; whilst with other species having fixed times for flowering, but 'which had been tempted by a mild January to put forth a few wretched flowers', no pollen was discharged from the anthers, and no seed was formed. The flowers of *Lysimachia vulgaris* if fully exposed to the sun expand properly, while those growing in shady ditches have smaller corollas which open only slightly; and these two forms graduate into one another in intermediate stations. Herr Bouché's observations are of especial interest, for he shows that both temperature and the amount of light affect the size of the corolla; and he gives measurements proving that with some plants the corolla is diminished by the increasing cold and / darkness of the changing season, whilst with others it is diminished by the increasing heat and light.[31]

The belief that the first step towards flowers being rendered cleistogamic was due to the conditions to which they were exposed, is supported by the fact of various plants belonging to this class either not producing their cleistogamic flowers under certain conditions, or, on the other hand, producing them to the complete exclusion of the perfect ones. Thus some species of Viola do not bear cleistogamic flowers when growing on the lowlands or in certain districts. Other plants when cultivated have failed to produce perfect flowers during several successive years; and this is the case with *Juncus bufonius* in its native land of Russia. Cleistogamic flowers are produced by some species late and by others early in the season; and this agrees with the view that the first step towards their development was due to climate; though the periods at which the two sorts of flowers now appear must

[31] For the statement by Linnaeus, see Mohl in *Bot. Zeitung*, 1863, p. 327. Asa Gray, *American Journal of Science*, 2nd series, vol. xxxix, 1865, p. 105. Bennett in *Nature*, November, 1869, p. 11. The Rev. G. Henslow also says (*Gardener's Chronicle*, 1877, p. 271: also *Nature*, 19 October, 1876, p. 543) 'that when the autumn draws on, and habitually in winter for such of our wild flowers as blossom at that season', the flowers are self-fertilized. On Lysimachia, H. Müller, *Nature*, September, 1873, p. 433. Bouché, *Sitzungsbericht der Gesell. Naturforsch. Freunde*, October, 1874, p. 90.

since have become much more distinctly defined. We do not know whether too low or too high a temperature or the amount of light acts in a direct manner on the size of the corolla, or indirectly, through the male organs being first affected. However this may be, if a plant were prevented either early or late in the season from fully expanding its corolla, with some reduction in its size, but with no less of the power of self-fertilization, then Natural Selection might well complete the work and / render it strictly cleistogamic. The various organs would also, it is probable, be modified by the peculiar conditions to which they are subjected within a completely closed flower; also by the principle of correlated growth, and by the tendency in all reduced organs finally to disappear. The result would be the production of cleistogamic flowers such as we now see them; and these are admittedly fitted to yield a copious supply of seed at a wonderfully small cost to the plant.

I will now sum up very briefly the chief conclusions which seem to follow from the observations given in this volume. Cleistogamic flowers afford, as just stated, an abundant supply of seeds with little expenditure; and we can hardly doubt that they have had their structure modified and degraded for this special purpose; perfect flowers being still almost always produced so as to allow of occasional cross-fertilization. Hermaphrodite plants have often been rendered monoecious, dioecious or polygamous; but as the separation of the sexes would have been injurious, had not pollen been already transported habitually by insects, or by the wind from flower to flower, we may assume that the process of separation did not commence and was not completed for the sake of the advantages to be gained from cross-fertilization. The sole motive for the separation of the sexes which occurs to me, is that the production of a great number of seeds might become superfluous to a plant under changed conditions of life; and it might then be highly beneficial to it that the same flower or the same individual should not have its vital powers taxed, under the struggle for life to which all organisms are subjected, by producing both pollen and seeds. With / respect to the plants belonging to the gyno-dioecious subclass, or those which co-exist as hermaphrodites and females, it has been proved that they yield a much larger supply of seeds than they would have done if they had all remained hermaphrodites; and we may feel sure from the large number of seeds produced by many plants that such production is often necessary or advantageous. It is therefore probable that the two forms in this subclass have been separated or developed for this special end.

Various hermaphrodite plants have become heterostyled, and now exist under two or three forms; and we may confidently believe that this has been effected in order that cross-fertilization should be assured. For the full and legitimate fertilization of these plants pollen from the one form must be applied to the stigma of another. If the sexual elements belonging to the same form are united the union is an illegitimate one and more or less sterile. With dimorphic species two illegitimate unions, and with trimorphic species twelve are possible. There is reason to believe that the sterility of these unions has not been specially acquired, but follows as an incidental result from the sexual elements of the two or three forms having been adapted to act on one another in a particular manner, so that any other kind of union is inefficient, like that between distinct species. Another and still more remarkable incidental result is that the seedlings from an illegitimate union are often dwarfed and more or less or completely barren, like hybrids from the union of two widely distinct species. /

INDEX

Printed and bound by CPI Group (UK) Ltd, Croydon, CR0 4YY

23/10/2024

01777667-0012